New Economic Windows

Domenico Delli Gatti · Edoardo Gaffeo
Mauro Gallegati · Gianfranco Giulioni
Antonio Palestrini

Emergent Macroeconomics

An Agent-Based Approach to Business Fluctuations

Springer

Domenico Delli Gatti
Catholic University of Milan, Italy

Edoardo Gaffeo
University of Trento, Italy

Mauro Gallegati
Polytechnic University of Marche, Ancona, Italy

Gianfranco Giulioni
University "G. d'Annunzio", Pescara, Italy

Antonio Palestrini
University of Teramo, Italy

Library of Congress Control Number: 2008921390

ISBN 978-88-470-0724-6 Springer Milan Berlin Heidelberg New York
e-ISBN 978-88-470-0725-3

Springer is a part of Springer Science+Business Media

springer.com

Printed in Italy

Cover design: Simona Colombo, Milano
Cover figure: © www.pixed2000.org
Typesetting: LE-TEX Jelonek, Schmidt & Vöckler GbR, Leipzig, Germany
Printing and binding: Grafiche Porpora, Segrate (MI)

Springer-Verlag Italia – Via Decembrio 28 – 20137 Milano

Printed on acid-free paper

To Davide, Marta and Tommaso,
Silvia and Giacomo, Sofia and Vittoria, and Sara
whose smooth individual behaviour emerges into chaos
when there is interaction.

Foreword and Acknowledgements

This book has grown out of the long-lasting interaction among researchers with heterogeneous skills and sensibilities in the group which promoted WEHIA (Workshop on Economies with Heterogeneous Interacting Agents) at the University of Ancona in 1996. The success of that initiative has been amazing. Ten years ago we did not expect so many people from so many countries to be so eager to discuss their work. With the benefit of hindsight, we can now detect an underground need to compare methodologies, conceptual and analytical frameworks in an exciting new field. At present, WEHIA is a well established international forum for discussion and cross fertilization of ideas on social interaction of heterogeneous agents.

Our starting point was (and still is) a deep dissatisfaction with the Representative Agent approach to macroeconomics and the companion idea that agents interact only through an anonymous market signal such as the price vector. In our opinion, there must be something wrong with a science which encounters embarrassing difficulties in explaining in a convincing way crucial phenomena such as the origin of money, the reasons for unemployment, the role of banks – to name only a few – and recurs to calibration of the model parameters to fit the empirical evidence. Suffice it to note, *en passant*, that had this practice of scientific discovery been used by astronomers during the last five centuries, we would still believe in the Ptolemaic system as the guiding principle for spatial explorations.

Going back to economics, if interactions and aggregations are ruled out from the beginning of the analysis, there will be no substantial difference between microeconomics and macroeconomics. As any bright student easily recognizes, the only remaining difference is that micro is taught on Monday and Tuesday and macro on Thursday and Friday (well, Wednesday is devoted to econometrics).

Economists have always been fond of the idea of the invisible hand governing the efficient allocation resources in a market economy. Alas, the Walrasian Auctioneer, i.e. the metaphor employed to model decentralized decision making, implies that equilibrium prices are determined through a centralized market clearing mechanism. The Walrasian approach abstracts from the way in which real-world transactions take place. By construction, interactions among agents are ruled out with the only exception of the indirect interaction through a clearinghouse institution.

After years of haunting with scientists exploring complex systems we are convinced that direct and/or indirect interaction among heterogeneous agents at the microeconomic level is a sufficient condition for macroeconomic regularities to emerge. Moreover, the interaction of microeconomic behaviour based on rules of thumb of a multitude of dispersed individuals can develop into some form of aggregate rationality. The main idea which percolates through this book is that aggregate phenomena (i.e. the dynamics of gross domestic product, the general price level etc.) cannot often be inferred from the behavior of the Representative Agent in market equilibrium continuously brought about by the implicit coordination of the Walrasian auctioneer.

On the contrary, aggregate phenomena emerge spontaneously from the interactions of individuals struggling to coordinate their actions on markets: macroscopic regularities emerge from microscopic behaviour. In other words, aggregate "laws" are due to emergence rather than to microscopic rules. In turn, *emergent macroeconomic dynamics* feeds back on microeconomic behavior through a downward causation process, in which economic and social structures affect the evolution of opportunities and preferences characterizing microeconomic units.

Mainstream, axiomatic economics is right: the invisible hand is often truly invisible. It continues to be out of sight simply because it is of a completely different nature than we were used to think so far or it has never been where it has been looked for.

The list of people who deserve our thanks for the help they provided during the preparation of this book is very long. We owe a huge intellectual debt to Alan Kirman, Joe Stiglitz, and to numerous participants of various WEHIA conferences, *in primis* Masanao Aoki and Thomas Lux, who have all been very inspiring. Special thanks to Beppe Grillo, a comedian with a penchant for economic analysis whose unorthodox view of the economy is surprisingly insightful. The vision outlined in this work has been refined in the course of stimulating conversations with many friends, in particular Bob Axtell, Xavier Gabaix and Matteo Marsili. It is that all of them are still friends even after having paid attention to our thoughts on the issue at hand. Of course, we are deeply indebted to many co-authors we had the opportunity to work with during the last twenty years or so (Anna Agliari, Tiziana Assenza, Tomaso Aste, Stefano Battiston, Carlo Bianchi, Michele Catalano, Pasquale Cirillo, Fabio Clementi, Giovanna Devetag, Marco Gallegati, Corrado Di Guilmi, Tiziana Di Matteo, Giorgio Fagiolo, Anna Florio, Yoshi Fujiwara, Laura Gardini, Bruce Greenwald, Nozomi Kichiji, Roberto Leombruni, Riccarda Longaretti, Mauro Napoletano, Paul Ormerod, Barkley Rosser, Alberto Russo, Emiliano Santoro, Enrico Scalas, Wataru Souma, Roberto Tamborini, Pietro Vagliasindi). All of them should be considered accomplices for the outcome you have in front of you. We also thank Simone

Alfarano, Gian-Italo Bischi, Buz Brock, Bruno Contini, Guido Fioretti, Cars Hommes, Paolo Pin, Peter Richmond, Leigt Testfasion, Richard E. Wagner, for their comments to an early version of this book. Comments and suggestions from participants in many conferences and workshops held at several Institutes and Universities (Bank of France; Bank of Italy; Unicredit Bank, Milan; Econophysics meetings at Bali, Canberra, Oxford and Warsaw; ISI Foundation, Turin; Lorenz Center, Leiden; Santa Fe Institute; Universities of Bielefield, Chieti-Pescara, Durham, Lille, Marseille, Milano "Bicocca", Milano "Cattolica", Pisa, Rome "La Sapienza", Sapporo, Seattle, Siena, Teramo, Tokio "Chuo", Trento, Udine) helped us very much. Finally, financial support from the MIUR (PRIN03 and FIRB02), the INFIM and the Università Cattolica di Milano, UPM and Trento is gratefully acknowledged.

The material assembled in this book is the outcome of a long-lasting endeavor. Our kids often complained about the time it took away from playing with them, asking "when it will be finished?" or firmly stating that they "can't stand it any more". We hope these same thoughts will not come to the mind of the reader while going through the book.

DDG, EG, MG, GG, AP

Table of Contents

1 Crucial Issues

"Economists study the actions of individuals,
but study them in relation to social rather than individual life"
Principles of Economics, A. Marshall

1.1 Introduction

The conceptual divide between microeconomics and macroeconomics is usually associated in textbooks to the different viewpoints from which the economy is looked at. While the focus of microeconomists is the study of how individual consumers, workers and firms behave, macroeconomics deals with national totals and, in doing that, any distinction among different goods, markets and agents is simply ignored. The methodological device to accomplish such a task is aggregation, that is the process of summing up market outcomes of individual entities to obtain economy-wide totals. However, what macroeconomists typically fail to realize is that the correct procedure of aggregation is not a sum whenever there exists interaction of heterogeneous individuals. Aggregation is therefore a crucial step: it is when *emergence* enters the drama. With the term *emergence* we mean the becoming of complex structures arising from simple individual rules (Smith, 1937; Hayek, 1948; Schelling, 1978). The physics taught us that to consider the whole as something more than its constitutive parts is a physical phenomena, not only a theory. Empirical evidence, as well as experimental tests, shows that aggregation generates regularities, i.e. quite simple and not hyper-rational individual rules when aggregated becomes well shaped: regularities emerge from individual "chaos". This book is a first, modest, step from the economics as an *axiomatic* discipline toward a *falsiable* science at micro, meso and macro level. It also tries to go into the details of economic interactions and their consequences for aggregate economic variables. By doing so, we suggest the agent based methodology as a framework for sound microfoundations of macroeconomics[1]. According to us, mainstream economics by ignoring interaction and emergence, commits what in philosophy is called "fallacy of division", i.e. to attribute properties

[1] For an other very interesting approach, discussing the social interaction framework to derive the evolution of macrovariables, see Brock-Durlauf (2005).

to a different level than where the property is observed (game theory offers a good case in point with the concept of Nash equilibrium, by assuming that social regularities come from the agent level equilibrium).

In particular, we are interested in applying this perspective to what is probably the most important single problem in macroeconomics: the analysis of the business cycle. We will do it in a untraditional way which differs from both the mainstream analysis (the impulse-propagation approach) and the disequilibrium approach, analyzing the business cycle as the outcome of the complex interaction of firms and industries (a procedure reminiscent of Schumpeter, 1939) in which small shock and endogenous elements coexist. In the physical jargon: individual behavior is at the root of the phenomenon, but when we aggregate or analyze the whole system a picture quite different from its constitutive elements emerges which allows to ignore the individual dynamics. In the following we will show that, even if this methodology is correct, we can keep track of the behavior of the aggregate and of the a very large quota of the individual firms at a very high confidence level.

From the very beginning of the discipline, the recurrence of upturns and downturns of aggregate output has fascinated the profession. In the period of time which spans from the end of World War I to the eve of the 21st Century, theoretical explanations of the business cycle have been loosely inscribed in two main contending methodological approaches. On the one hand, there is the so-called *impulse-propagation* or *equilibrium* approach, in which large exogenous stochastic perturbations are superimposed to a system of linear (or suitably linearized) deterministic difference/differential equations describing the dynamic relationships between economic variables.[2] Since, by assumption, the solution of the underlying deterministic system is unique and stable, expansions and contractions driven by random disturbances occur around a stable (general) equilibrium, while fluctuations themselves are stationary stochastic processes. There is nothing like a cycle, according to this definition but, rather, "recurrent fluctuations of output around trend and co-movements among other aggregative time series" (Kydland and Prescott, 1990). Interestingly enough, such an analytical device has been equally applied to explanations of the business cycle devised by competing schools of thought, suffice it here to cite the monetarist model of Lucas (1975), the real business cycle model of Kydland and Prescott (1982), or the New Keynesian model of Taylor (1980).

At the other end of the methodological spectrum one can find the *endogenous* approach to business cycles. This class of models, of which the most prominent ones are those by Kaldor (1940) and Goodwin (1967), does not rely on some external shock to account for business cycle phenomena. Instead, cycles are conceived of as self-sustained oscillations, a result obtained

[2] The idea of explaining the mathematical nature of business fluctuations in terms of a combination of deterministic and stochastic components can be traced back to the work of Frisch (1933) and Slutzky (1937).

by exploiting the disequilibrium and non-linear relationships among economic aggregates. From an empirical point of view, this approach resemblaces the old NBER view, according to which: "the business cycle [...] consists of expansions occurring at about the same time in many economic activities, followed by similairly general recessions, contractions, and revivals which merge into the expansion phase of the next cyxle" (Burns and Mitchell, 1946). They add that the movement, although recurrent, is not periodic, lasting from 1 to 12 years, and it is not divisible into shorter cycles.

Of course, both approaches are not free from limits and inconsistencies. In spite of the equilibrium approach having nowadays became the workhorse of modern macroeconomics, for example, their users still find enormous difficulties in explaining why small shocks produce large fluctuations. A well-known argument in multi-sector real business cycle models (see e.g. Long and Plosser, 1983) is that as the number of sectors or industries considered in the analysis becomes large, aggregate volatility must tend to zero very quickly. This result, which follows directly from the *Law of Large Numbers* (LLN), rests on the hypothesis that each sector is periodically buffeted with idiosyncratic, identically and independently distributed shocks to Total Factor Productivity (TFP). As negative and positive shocks tend to cancel out, in an economy consisting of N sectors – each one of approximately size $1/N$ of GDP – aggregate volatility must converge to zero at a rate $N^{1/2}$ (Lucas, 1981). Furthermore, under rather general conditions, such a curse of dimensionality is so compelling to offset any shock-propagation effects due to factor demand linkages among industries (Dupor, 1999). Hence, for a multi-sector neoclassical business cycle model to be able to replicate aggregate fluctuations with a degree of volatility in line with that observed in real data, one has necessarily to appeal to aggregate shocks (but the empirical evidence seems to rejects this hypothesis).

The disequilibrium approach, in turn, shares with its competing mate the major limitation of being completely time reversible. In such a case, the Laplace demon would be able to predict the future (or to re-construct the history) of a system by simply knowing the exact actual conditions. If such a hypothesis is accepted, then historical time is out of the game and reversibility (or *time reversal symmetry*, as the physicists define it) follows directly.

However, it seems to us that the most severe drawback of both approaches, and in turn of modern theorizing about macroeconomic fluctuations (growth theory, aggregate consumption, aggregate investment, and so on) as a whole, relates to the unsolved issue of the exact relationship between statements at the microeconomic level in terms of behavioral rules and aggregate categories, like income, expenditure or savings. The two issues at stake are, on the one hand, how to address the remarkable and persistent heterogeneity among individual economic entities, and, on the other hand, the fact that in real-world situations agents do not take their decisions in isolation but are influenced by the network of social affiliations whom they belong to.

1.2 Aggregate Among Peers – If You Please

Mainstream economics is based on *reductionism*, i.e. the practice of scientific discovery at the root of classical physics. On the one hand, the *ceteris paribus* method developed by Marshall reflects the idea of a physical world which can be suitably described by a dynamical system capturing some features of nature in isolation, and an environment which affects the object of study only by means of perturbations. On the other hand, economists generally accept that structures at an aggregate level can be deduced and predicted just by looking at the individual components of the system. The key principle which has guided neoclassical economics since its inception is the restricted idea of equilibrium as developed in rational mechanics (Mirowski, 1989), in particular in its static version attributed to Archimedes (McCauley, 2004). As should become clear, such a methodology of scientific advancement is likely to be successful in economics only if: a) the functional relationships among variables are linear, and b) there is no direct interaction among economic units.

If one "translates" these 2 conditions into economic terms, she actually assumes a very particular nature of the economic system: i) all the n-agents are connected to a single coordinating individual, an auctioneer or a planner; ii) all the information is freely mediated by this guy. In the most extreme case, any individual strategy is excluded and agents have to be uniform. Small departures from perfect information open up the chance of having direct links, thus changing the economic network and therefore violating conditions a–b).

Refusal of conditions a) and b) are the two minimum requirements to define a complex system. What characterizes complex system is the notion of emergence, that is the spontaneous formation of self-organized structures at different layers of a hierarchical system configuration (Crutchfield, 1994).

Since economies are complex systems and non-linearities are pervasive, mainstream economics generally adopts the trick of linearizing functional relationships. Moreover agents are supposed to be all alike and not to interact. Therefore, any economic system can be conceptualized as consisting of several identical and isolated components, each one being a copy of a *Representative Agent* (RA). The aggregate solution can thus be obtained by means of a simple "N-replication" of the choices made by each optimizing agent.

The RA device, of course, is a way of avoiding the problem of aggregation by eliminating heterogeneity and, in some situations, such a simplification may capture real world qualitative features. But heterogeneity is still there. If the macroeconomist takes it seriously, he/she has to derive aggregate quantities and their relationships from the analysis of the micro-behaviour of different agents. This is exactly the key point of the *aggregation problem*: starting from the *micro-equations* describing/representing the (optimal) choices of the economic units, what can we say about the *macro-equations*? Do they have the same functional form of the micro-equations (the *analogy principle*)? If not, how to derive the macro-theory?

The aggregation problem in macroeconomics has a long history. Since Gorman (1953) it is well known that the conditions for exact aggregation are stringent and almost never satisfied. Stoker has gone so far as to propose a methodology for stochastic aggregation. Aoki (1996, 1998) has put forward a combinatorial method. These efforts are welcome but the science (or the art?) of aggregation is still in its infancy.

1.3 Robinson Crusoe Meets Friday

A distinctive feature of the (nowadays mainstream) neoclassical school of thought is the chase for sound microfoundations for macroeconomic analysis, as a methodological overtaking of the Keynesian approach centred on aggregate categories. Conceptually, a description of how this endeavour has been substantiated in theoretical modelling requires two steps. The first one consists in assuming that all "[...] *relative prices are determined by the solution of a system of Walrasian equation*" (Friedman, 1968, p. 3), in order to apply such a framework with brute force to a macroeconomic problem. No attention is paid to the well-known fact that the Walrasian general equilibrium model does not guarantee either the stability or the uniqueness of the general equilibrium itself (Kirman, 1989) or to who does change the price in a perfect competition setting (Arrow, 1959). The second step derives consequently from the first one, and we find no better way to express it than to recur to the following quotation from Plosser (1989, p. 55):

> "*How does one think about the competitive equilibrium prices and quantities that are implied by this framework? The first step is to recognize that all individuals are alike, thus it is easy to imagine a representative agent, Robinson Crusoe, and ask how his optimal choices of consumption, work effort and investment evolve over time.* [...] *(W)e can interpret the utility maximizing choices of consumption, investment and work effort by Robinson Crusoe as the per capita outcomes of a competitive economy*".

No caveats! The simplifying hypothesis of a RA might be far too simplifying, but it is instrumental for greed, rationality and equilibrium to be the only necessary and sufficient conditions for scientific macroeconomic theory. Standard economics is not falsifable since it became an axiomatic discipline.

Admittedly, some dissenting voices urging towards an analysis of how social relations affect the allocation of resources resounded loudly from the start even in the rooms of the neoclassical citadel (e.g., Leibenstein, 1950; Arrow, 1971; Pollack, 1975). They went almost completely unheard, however, until the upsurge in the early 1990s of a brand new body of work aimed at understanding and modeling the social context of economic decisions, usually labeled *new social economics* or *social interaction economics* (Durlauf and Young, 2001).

The key idea consists in recognizing that the social relations in which individual economic agents are embedded can have a large impact on economic decisions. In fact, the social context impacts on individual economic decisions through several mechanisms. First, social norms, cultural processes and socio-economic institutions may influence motivations, values, tastes and, ultimately, make preferences endogenous (Bowles, 1998). Second, even if we admit that individuals are endowed with exogenously-given preferences, the pervasiveness of information asymmetries in real-world economies implies that economic agents voluntarily share values, notions of acceptable behavior and socially-based enforcement mechanisms in order to reduce uncertainty and favor coordination (Denzau and North, 1994). Third, the welfare of individuals may depend on some social characteristics like honor, popularity, stigma or status (Cole *et al.*, 1992). Finally, interactions not mediated by enforceable contracts may occur because of pure technological externalities in network industries (Shy, 2001) or indirect effects transmitted through prices (pecuniary externalities) in non-competitive markets (Blanchard and Kyiotaki, 1987), which may lead to coordination failures due to strategic complementarities (Cooper, 1999).

A useful operational classification of the channels through which the actions of one agent may affect those of other agents within a reference group is given by Manski (2000), who distinguish among: i) *constraint interactions*: the decision to buy or sell by one agent influences the price of the good, thus affecting the feasible choice set of other individuals; ii) *expectations interactions*: asymmetric information on markets means that one agent forming expectations of the future course of relevant variables may try to augment his/her information set by observing the actions chosen by others (*observational learning*), under the assumption that this could reveal private information; iii) *preference interactions*: the preference ordering over the choice set of one agent depends directly on the actions chosen by other agents.

Models of social interactions are generally able to produce several interesting properties, such as *multiple equilibria*, when the social component of utility (e.g., the social pressure to conform to the average education level) is higher than the private one (e.g., private expected return to education) (Brock and Durlauf, 2001); *non-ergodicity* due to the path-dependency feature of the statistical equilibrium and *phase transition*, that is the passage from a state of multiplicity of equilibria to one of uniqueness, at a critical threshold ratio between private and social utility (Durlauf, 1993); a tendency toward *equilibrium stratification* in social and/or spatial dimension (Benabou, 1996; Glaeser *et al.*, 1996); and finally the existence of a *social multiplier* of behaviors (Glaeser *et al.*, 2002).

The stage is now complete for presenting the main message in this book: heterogeneity matters, interactions amplify its role in shaping aggregate responses and the economic regularities emerge from the interaction of heterogeneous agents. Interaction and adjustment involve dynamics at the individual level and, as Axtell (2001) shows, is not a fixed point (it is complex). Macroscopic regularities emerge from the interactions of the agents: microequilibrium is sufficient to have macroequilbrium, but it is not necessary at agent based level, where there are fluctuations, continuous adaptation and adjustement to one another: here is the room for computation, or the agent based modeling we develop in Chap. 3.

1.4 Complexity

Complexity is a complex word: It has a lot of heterogeneous interacting meanings. For simplicity, we can focus on two (among many) ways in which the qualifier "complex" has shown up in economics. In the literature of the '80s, it has been associated with dynamics: The expression *complex dynamics* has been often used as a synonym of *chaos* or chaotic dynamics. As such it has been essentially applied to the evolution over time of macro variables. Chaotic motion, in fact, is characterized by the simultaneous properties of local instability and global stability that is so attractive for business cycle theorists. Since Goodwin and the introduction of limit cycles in macroeconomics, in fact, the idea of the macroeconomy being locally repelled by an unstable state and globally converging to a cycle has been an intriguing feature of macrodynamics.

Limit cycles, however, are "too regular". Chaotic dynamical structures are even more appropriate for business cycle analysis because of the deterministic unpredictability of the time series they generate. There are plenty of models in this line of research (a pioneer in this field is R. Day, see, e.g. Day, 1994). Most of these models are aggregative in nature: By construction, they do not deal with the microeconomic components or determinants of macrovariables. In any case there is not much reflection in this literature on the relationship between individual and macroeconomic behaviour. Attention is paid mainly to the "irregular" (complex) – i.e. aperiodic and asymmetric – dynamics of the time series generated by sometimes very simple non linear mathematical structures, which are observationally equivalent to those generated by a linear structure continuously affected by stochastic disturbances.

This line of research is still active but somehow less thriving. There are many reasons for this. First: Results are not robust for deterministic systems[3]. Very small changes in the parameters yield huge qualitative changes in the properties of the dynamics generated by these models. Second: even though chaotic regions may have positive Lebesgue measure in the parameter space, complex dynamics occur often for particular, generally small, intervals of the parameters of interest. Third: From the empirical point of view, it is

[3] If we add noise to the system then there may be "robust features" determined by the underlying invariant measure such as the autocorrelation pattern of noisy chaotic time series (as an example see Hommes, 1996).

extremely difficult to find traces of complex deterministic dynamics in the macroeconomic time series. The econometric tests developed to suit this purpose, such as the BDS test, are only capable to discern non-linearity in the structure of the economy but do not detect the particular type of non-linearity which is necessary for chaotic dynamics[4].

With the passing of time, the meaning has slowly shifted so that the qualifier complex is now usually associated with the working of economic structures with heterogeneous interacting agents. The so-called science of complexity, which has grown out of the joint efforts of hard and soft scientists in the '80s and '90s, in fact, claims that there are common properties of *complex systems* which are the object of study in many different fields such as the cell, the brain, language, the capitalist market economy. The focus therefore has moved from the macro to the micro level. Macroeconomic variables can be reconstructed by summing up individual magnitudes (bottom up procedure). Complex structures consisting of heterogeneous interacting agents generate complex dynamics also of the macrovariables (a case in point is the model of Chap. 3).

Complex economic structures are usually associated with adaptive agents so that they are often referred to in the literature as Complex Adaptive Systems. The pre-analytic vision of a complex market economy, in fact, is centered upon agents endowed with limited information and computational capability (bounded rationality) so that they adopt rules of thumb (instead of optimization procedures) and are naturally led to interact with other agents to access information, learn and imitate. In this sense, complexity goes hand in hand with evolutionary dynamics and direct interaction among agents. Complex structures, however, can emerge even when agents are rational, i.e. they maximize an objective function subject to constraints and interaction is only indirect (once again a case in point is the model in Chap. 3).

Sometimes complexity applied to economics overlaps with *econophysics*. The underlying methodological assumption of econophysics is that, even if economics is a social science and has to deal with incentives and human decisions the aggregate behaviour can be described by models of statistical physics. Collective behaviour is the outcome of the interaction of many heterogeneous individuals in ways which recall the interaction of particles in statistical mechanics. Recent works in econonophysics has focused mainly on three issues: the analysis of the time series of Stock prices, exchange rates and goods prices; the evolution over time of the distribution of firms' size, individual wealth and income; the exploration of economic phenomena by means of networks. In this book we contribute to the second strand of literature.

---

[4] In case of "chaos plus noise" a recent literature do not reject the possibility of chaos buffered by small dynamic noise (see Hommes-Manzan, 2006).

1.5 Outline of the Book

In this book we explore the consequences of heterogeneity of firms' size and degree of financial fragility in a financial accelerator model along the lines of Greenwald and Stiglitz (1993). The presence of imperfect information has several important consequences in the modeling strategy. First of all, agents have to be heterogeneous, since the access to different informative sets discriminate among them. Then, agents are ex-post bounded rational, since their constrained information prevent them to be in a Pareto efficient, even if sub-optimal, equilibrium position (Greenwald and Stiglitz, 1990). Finally, if agents are heterogeneous they interact outside the price system while time becomes important since future is uncertain. A decision to produce today will affect the future because of debt commitment (which depends on the firm's past decision as well from the behavior of the other firms) is a crucial issue for the future profits. A financially fragile framework should encompass it.

Before going into the detail of the model, in Chaps. 2 and 3 we present and discuss the empirical evidence on the evolution over time of the distribution of relevant industrial variables. In Chap. 4, we explore the link between the distributions of firms' size and rate of growth, showing that the power law distribution of firms' size may be at the root of the fat-tail distribution (often approximated, over a relevant subset of the support, by a double exponential or Laplace-type distribution) of the growth rates. Our main claim is that the evolution over time of these distributions is of central importance not only in industrial organization but also in business cycle analysis.

In Chap. 4 we also present the model. In our approach, the origin of fluctuations can be traced back to the everchanging configuration of the network of heterogeneous firms. A major role in shaping dynamics is played by financial variables. In the absence of forward markets, the structure of sequential timing in our economy implies that agents have to rely on credit to bridge the gap between decision and realization. Highly leveraged – i.e. financially fragile – firms, are exposed to the risk of default. When bankruptcies occur, non performing loans affect the net worth of the banking system, which reacts reducing the supply of credit. Shrinking credit supply makes interest rates go up for each and every firm increasing the risk of bankruptcy economywide. A snowball effect consisting in an avalanche of bankruptcies can follow.

Chapter 5 is devoted to the discussion of further issues in this line of research. It concludes the book but is not a conclusion at all. The research project we would like to carry out is still in its inflacy. The present book cover a non negligible but still short distance in what we think is the right direction.

2 Stylized Facts of Industrial Dynamics: The Distribution of Firms' Size

2.1 Introduction

Economists interested in business cycle and growth theory have long been trained to the use of *stylized facts* as a practical guide in implementing their research agenda, as the pioneering accounts of Burns and Mitchell (1946) and Kaldor (1963) testify. The advent of the neo-classical counter-revolution in the late 1960s, rooted in what Robert Solow dubbed the holy trinity of Rationality, Equilibrium and Greed, has somehow inverted the logic of scientific discovery in economics. The first step in nowadays orthodox macroeconomics consists in building a model of microeconomic behavior based on axiomatic descriptions of preferences and technology. Afterwards, the model is solved *via* the representative agent and taken to the data. Alas, as shown *inter alia* in Caballero (1992) and Kirman (1992), the falsifiability of the model may be fatally prevented due to a *fallacy of composition*, that is the presumption that what is true of each single part of a whole is necessarily true of the whole as well. In particular, the straight application of a microeconomic rationale to aggregate data can be seriously misleading whenever the probabilistic forces at work as the number of entities grow large, i.e. the Central Limit Theorem, the Law of Large Number or any of their extensions, are not properly taken into account.

A proper methodology to tackle these issues consists of two pillars. First, empirical laws at a macroeconomic level should be expressed in terms of statistical distributions, such as the distribution of people according to their income or wealth, or the distribution of firms according to their size or growth rate (Steindl, 1965). A great deal of useful information and several additional questions waiting for a scientific explanation can be derived by looking at such empirical distributions and their invariant, or long-term, character as the cumulated responses of individual entities concerning their choices of labor supply, investment demand, pricing, and so on. Second, suitable modeling strategies should be adopted, that is explanatory methodologies capable to combine a proper analysis of the behavioral characteristics of individual agents and the aggregate properties of social and economic structures (Sunder, 2006).

Anecdotic and econometric evidence largely confirm the coexistence of firms and households characterized by non negligible and persistent heterogeneity along several dimensions (Haltiwanger, 1997; Diaz-Gimenez *et al.*, 1997). For example, it is well known at least since Gibrat (1931) that the size distribution of firms is right skewed for several different countries and historical periods (De Wit, 2005). A more recent stylized fact on firm dynamics concerns the distribution of firms' growth rates, which appears to be approximated by a Laplace (double-exponential) distribution (Amaral *et al.*, 1997; Bottazzi *et al.*, 2002). Finally, earnings, income and wealth are well-known to be highly concentrated over households, regardless of the measure of concentration, i.e. the Gini coefficient, the coefficient of variation or the inter-quartile ratio (Diaz-Gimenez, 1997). What is missing is an analysis of the interrelationships between heterogeneity, its change and macroeconomic dynamics both in terms of business fluctuations and long-run growth.

In what follows we will add some new evidence on the shape of the distribution of heterogeneous firms and households, making use of several databases. In particular, we will focus on four issues: i) the shape of the long-run firms' size and growth rate distribution (Sect. 2.2); ii) the distribution of firms as they exit the market (Chap. 3); iii) the heterogeneity of productivity over firms, and of income over households; iv) the distributional features of business cycle phases. The last point is a first attempt to close the gap between micro and macro emphasized above.

2.2 Pareto, Gibrat, Laplace: The Statistical Analysis of Industrial Dynamics

The distribution of firms' size is empirically approximated by a Zipf or power law. A well known object mainly in physics and biology, the power law distribution has been originally derived more than a hundred years ago by Vilfredo Pareto, who argued that the distribution of personal incomes above a certain threshold follows a heavy-tailed distribution (Pareto, 1897). This fact baffled scholars since the Central Limit Theorem implies that the income distribution should be lognormal under the reasonable assumption that the rates of growth of income brackets are only moderately correlated.

A similar conundrum recurred again about 30 years later in industrial economics, due to the pioneering work of Gibrat who put forward the *Law of Proportional Effects* or *Gibrat's Law* (Gibrat, 1931). According to Gibrat's law in weak form, the growth rate of each firm is independent of its size. If the law of proportional effect is true, the distribution of firms' size will be right skewed. Gibrat went even further, arguing that, if the rates of growth are only moderately correlated, such distribution will be a member of the log-normal family (Gibrat's law in strong form).

In a nutshell, the size (measured by output, the capital stock or the number of employed workers) of the i-th firm K_{iT} in period T is defined as $K_{iT} = K_{iT-1}(1 + g_{iT})$, where g_{iT} is the rate of growth. Taking the log of both sides and solving back recursively from time 0 size K_{i0}, it is straightforward to obtain[1] $\log K_{iT} \cong \sum_{t=1}^{T} g_{it} + \log K_{i0}$. Assuming that the growth rates are identically independently distributed, the distribution of the log of firms' size tends asymptotically – i.e. for t approaching infinity – to the lognormal distribution. The reason is that under the central limit theorems assumptions one would expect that $\sum_{t=1}^{T} g_{it}$ tends to be normal.

Recent research has shed several doubts, however, both on the true nature of this stylized fact, and on its explanation. From a theoretical perspective, for example, it has been argued that stories based on pure random processes have too little economic content to be acceptable (Sutton, 1999). The empirical literature, on its part, has shown that attempts to make generalizations on the shape of size distributions for firms have generally failed (Schmalensee, 1989).

As a matter of example, in recent work Robert L. Axtell (2001) disputes the finding of log-normality for the size distribution of U.S. firms reported in Stanley *et al.* (1995), claiming that correct results should be expected only after recognizing the right proxy for firm sizes, and after adopting a sufficiently large sample. In particular, he finds that a Zipf or power law (or Pareto) distribution returns a good fit to the empirical one, and that the scaling exponent is strikingly close to 1 over time, a result which is partly consistent with early findings reported in Ijiri and Simon (1977).[2] Moreover, Stanley *et al.* (1996) and Amaral *et al.* (1997) have found that the growth rate of firms' output is better approximated by a Laplace distribution compared to a normal distribution.

To explain these facts, the literature has followed two lines of research. The first one is a-theoretical and focuses only on the statistical properties of the link between the distribution of the state variable (firms' size) and that of the rates of change. For instance, Reed (2001) shows that independent rates of change do not generate a lognormal distribution of firms' size if the time of observation of firms' variables is not deterministic but is itself a random variable following approximately an exponential distribution. In this case, even if Gibrat's law holds at the individual level, firms' variables will converge to a double Pareto distribution.

The second line of research – to which the model described in Chap. 4 belongs – stresses the importance of non-price interactions among firms hit by multiplicative shocks, hence building on the framework put forward by

[1] Taking the logarithm on both side and using the fact that $\log(1+g)$ is equal to $g + o(g)$ when g is small.

[2] To be precise, *Zipf's law* is the discrete counterpart of the Pareto continuous distribution (power law). It links the probability to observe the dimension of a social or natural phenomenon (firms' size, cities, earthquakes, words in a text, etc.) with rank greater than, say, κ to the complementary cumulative frequency. In case of firms' size the scale parameter is equal to 1. In words: the probability that the i-th firm has size Kit greater of equal to a certain level k is equal to $1/k$. In symbols: $\Pr(K_{it} \geq \kappa) \propto \kappa^{-1}$.

Herbert Simon and his co-authors during the 1950s and '60s (Ijiri and Simon, 1977). As a matter of example, Bottazzi and Secchi (2003) obtain a Laplace distribution of firms' growth rates within Simon's model, just relaxing the assumption of independence of firms' growth rates. In their model, using a finite linear Polya urn, they assume that if firm i had in the past k times opportunity to growth compared to firm j, than firm i has in the next period k times the probability of firm j to get a new opportunity to growth.

In principle these results can induce the reader to reject the strong version of Gibrat's law. After all, this law claims that the distribution of the levels (firms' size) is lognormal while the empirical analysis points to Zipf's law and the distribution of growth rates seems to be Laplace. As a matter of fact, things are not that simple. The idea according to which Gibrat's law has to be fully discarded is wrong, since in the recent literature a weak version seems to hold, in which growth rates seem to be independent at least in mean. In fact, Lee *et al.* (1998) show that the variance of growth rates depends negatively on firm's size. The implications of the strong version of Gibrat's law are not necessarily true in the weak version.

Critics to the scaling empirical evidence can be found in Quandt (1966) and Kwoka (1982) since they found systematic departures from the power law distribution at the sector level. A recent work shows that these findings (power law at the aggregate level and a plethora of distributions at the sector one) are consistent if firms' growth is characterized by common components (Axtell *et al.*, 2006).

2.3 Unconditional Firms' Size Distribution for Pooled International Data

Axtell (2001) puts forward the testable conjecture that the Zipf distribution may be the best fit of the empirical distribution of firms' size not only in the U.S. but also in other countries and calls for new evidence to be gathered and explored by means of alternative data-sets.

The evidence available so far on Axtell's conjecture is mixed.[3] Thus, it seems worthwhile to further extend the empirical analysis on cross-country samples, in order to test Axtell's null hypothesis of a Zipf distribution for firms' size outside the U.S., against the alternatives of a power law with scaling exponent different from 1 on the one hand, or of a log-normal distribution, on the other hand. First of all we analyze company account data extracted from the commercially available *Datastream International* (DI) data-set, which reports annual time series of company accounts for a sample of quoted companies. We focus on non-financial firms located in the G7 countries over the 1987–2000 time span.

[3] See, for example, Takayasu and Okuyama (1998), Voit (2000) and Knudsen (2001).

For the sake of pooling consistency, several selection criteria have been applied sequentially to obtain the final sample used for estimation. First, we removed from the sample firms with missing data points. Second, in order to control for the impact of major mergers and acquisitions, we excluded firms whose capital stock had changed by a factor of two or more from any one year to the next. Third, to avoid biased estimates due to outliers we removed firms with data points outside a conventional three standard deviations confidence band for any of the variables of interest.

Thus, the resulting panel is unbalanced both because there is a different number of observations for different firms, and because these observations may correspond to different points in time. The number of firms is 126 for Canada (1099 observations), 178 for France (1415 observations), 176 for Germany (1378 observations), 84 for Italy (460 observations), 748 for Japan (7416 observations), 376 for UK (3467 observations) and 682 for USA (6829 observations).

The variables we employed are total *sales*, y, total *capital employed*, k, and total debt, d. The last variable is not a conventional size variable. It seems however to be generally highly correlated with firms' dimension, so that we used it as an additional proxy for size. In order to pool all the countries together, we standardized all the variables by dividing them for their mean, so that one obtains quantities which are independent of the different account standards adopted in the G7 countries.

Roughly speaking, a discrete random variable Z is said to follow a Pareto-Levy (also known as Rank-Size or power law) distribution, if its complementary cumulative distribution function takes the form:

$$\Pr\left[Z \geq z_i\right] = \left(\frac{z_0}{z_i}\right)^{\alpha} \tag{2.1}$$

with $z_i \geq z_0$. The scaling exponent $\alpha > 0$ is also known as the shape parameter, while z_0 (the minimum size) is the scale parameter. On a log-log space, this distribution yields a downward sloping straight line with slope $-\alpha$. The special case $\alpha = 1$ is known as Zipf's Law.

In panels a) to c) of Fig. 2.1, we present the *log-log* plot of the frequency distribution of firms' size[4] for firms' real sales, total capital and debt, respectively. The interpolating line, which informs us on the goodness of fit of a Power Law distribution, has been determined by means of the following OLS regression:

$$\ln(f(S_i)) = a - (\alpha + 1)\ln(S_i) , \tag{2.2}$$

where S_i stands for firms' size, and $f(S_i)$ is the correspondent frequency, with $i = y, k, d$.

[4] The graph are computed using simple histograms. In order to avoid the bias caused by the small number of observations in the right tail, only frequencies bigger than 0.05% are displayed.

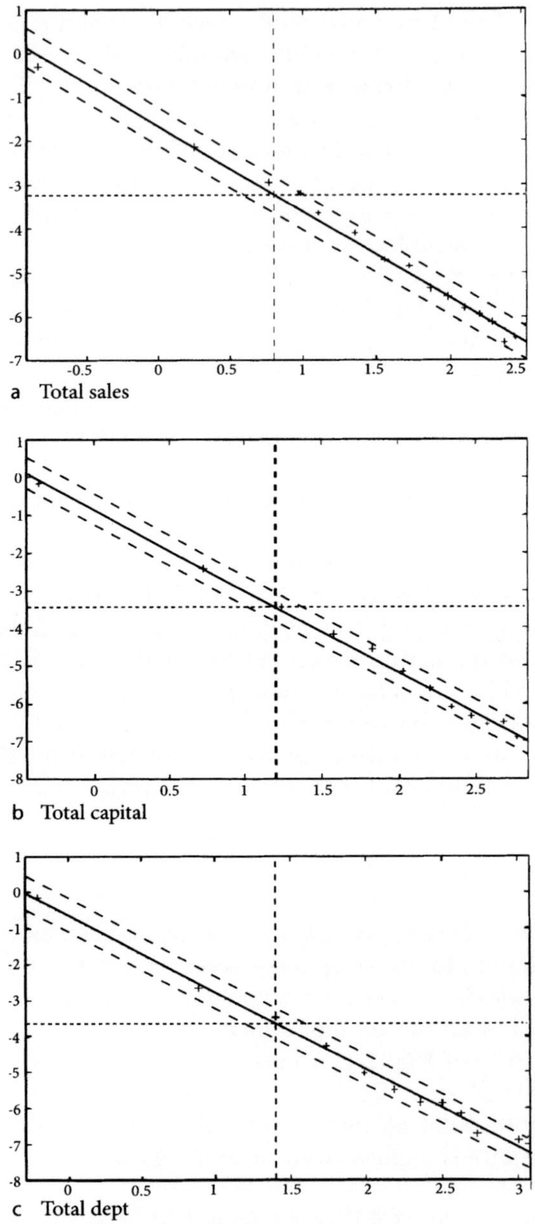

Fig. 2.1. Zipf plots of total sales (y), total capital employed (k) and total loan capital (d). The two *dashed lines* identify the 95% confidence interval for predictions

The point estimates of α are equal to 0.96 (34.53) for sales, to 1.16 (46.27) for capital, and to 1.14 (35.32) for debt, where figures in brackets represent t-values[5]. The goodness of fit is in any case truly remarkable, with values for adjusted R^2 equal to 0.979, 0.988 and 0.978, respectively.

Notice that only when the size is measured by total sales our findings are fully consistent with those in Axtell (2001) for the U.S., where size is measured in turn as the number of employees per firm. In fact, in this case the null hypothesis that the size distribution is Zipf, that is that the "true" α is 1, could not be rejected as the estimate returns a value which lays in a one standard deviation confidence band. If size is measured by means of capital or debt, instead, the distribution appears to be less even-sized than predicted by the Zipf Law.

2.4 The Size Distribution of Firms Conditional on the Business Cycle

As argued extensively in Brock (1999), the good linear fit of a distribution in the log-log space should be interpreted with great care, since these distributions are unconditional objects and many conditional data generating processes (DGPs) are consistent with them. Thus, in order to refine the evidence in a way which could be suitably used to discipline theory we condition the processes under scrutiny on business cycle episodes.

In other terms, we are interested in assessing whether the statistical models driving firms' growth change from upturns to downturns or, in other terms, whether firms long-run growth processes are influenced by short-run fluctuations.

To analyze this issue, we applied the Hodrick-Prescott (HP) filter[6] to the time series of the industrial production index for each country in order to detect country-specific recessions and recoveries. Recessions (recoveries) are then defined as the period between a peak (trough) and a trough (peak) in detrended industrial production, where a peak is the year before the de-trended index turned negative, and a trough as the year before it turned positive. Table 2.1 reports the ratio – which we label ry – of the estimated conditional mean of y calculated for recoveries to the estimated conditional mean of the same variable for recessions. The labels rd and rk are self-explaining. Moreover, in the same table we report the ratio – which we label $r\sigma(y)$ – of the standard deviation of y for recoveries to the standard deviation of the same variable for recessions. The labels $r\sigma(d)$ and $r\sigma(k)$ are self-explaining.

[5] Of course, t-values are referred to the "reduced form" parameter $(\alpha + 1)$.

[6] From a technical viewpoint, the HP is a low-pass filter. Hence, the cyclical component is obtained by subtracting from the raw series the filtered one. The smoothing parameter λ has been tuned at the value 100 for annual data. See Hodrick and Prescott (1997).

Table 2.1. Percentage ratio ri ($i = y, d, k$) of the estimated conditional mean of variable i in recoveries to the estimated conditional mean of the same variable in recessions and percentage ratio $r\sigma(i)$ of the standard deviation of variable i in recoveries to the standard deviation of the same variable in recessions

Country	ry	$r\sigma(y)$	rd	$r\sigma(d)$	rk	$r\sigma(k)$
Canada	−4.3	−12.6	1.7	−0.5	−10.9	−26.7
France	18.2*	29.6**	27.1*	61.2**	26.2**	57.6**
Germany	10.5	16.3**	11.5	4.7	9.1	10.7**
Italy	13.6	14.3*	19.2	22.1**	14.9	14.8*
Japan	1.0	2.3	−6.2	−2.3	−2.1	0.4
UK	17.3*	18.2**	43.4**	56.1**	28.9*	109.1**
USA	28.1**	27.3**	26.0*	32.4**	30.1**	42.5**
Joint test	18.66**	220.36**	20.215**	543.25**	20.451**	1128.3**

* denotes rejection of the null of no difference between expansion and recession *vs.* the alternative of bigger values in expansion, at the 5% significance level.

** denotes rejection of the null of no difference between expansion and recession *vs.* the alternative of bigger values in expansion, at the 1% significance level.

In other words, what these numbers say is how big are the above mentioned descriptive statistics in expansions relative to the magnitude they assume in recessions.

All the countries, with the exceptions of Canada (for sales and capital) and Japan (for debt and capital), present a common pattern: both the mean and the standard deviation of firms' size are bigger during expansions.[7] This effect is particularly important in the U.K. and the U.S. In the former country both the mean and the standard deviation of d increase of about 50% on average during expansions, while the standard deviation of k doubles. In the U.S. the huge increase in total sales during expansions suggests the presence of some kind of leverage effect. Overall, these figures suggest that there are significant changes in firms' distribution during the different phases of the business cycle. Indeed, as reported in the last row of Table 2.1 a χ^2 test distributed with 7 degrees of freedom rejects the null of no differences at the 1% significance level for the mean of each variable and its standard deviation.

In spite of this variability, it is particularly interesting to note from panels a) to c) of Fig. 2.2 that a power law scaling behavior emerges as an invariant feature of the size distribution of firms, regardless of the proxy used to measure size or of the phase of the business cycle used to condition the distribution.

Point estimates are reported in Table 2.2, from which it is clear that the linear fit is very good, but that only in two cases the conjecture of a Zipf distribution can not be rejected, namely for total sales during expansions

[7] As a matter of fact, the ratios for Canada and Japan result always statistically non-significant.

Table 2.2. Estimated scaling parameters and goodness of fit for linear log-log regressions. Numbers into brackets are t-values

	Expansions		Recessions	
	α	$\bar{R}^2$	α	$\bar{R}^2$
y	0.97 (28.25)	0.971	0.81 (30.90)	0.972
k	1.18 (31.82)	0.984	1.04 (27.43)	0.965
d	0.84 (34.00)	0.982	0.73 (23.58)	0.970

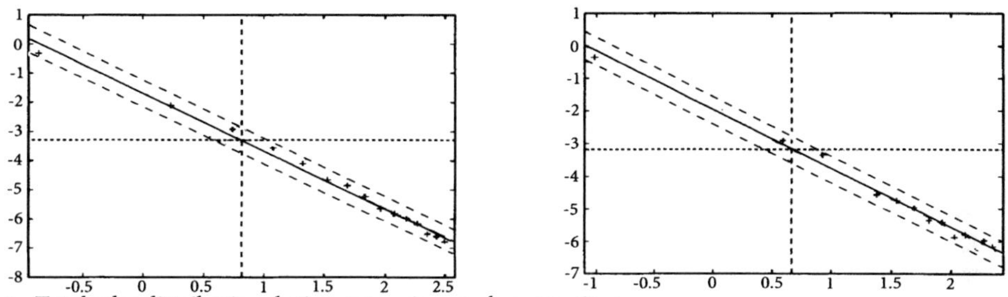

a Total sales distribution during expansions and contractions

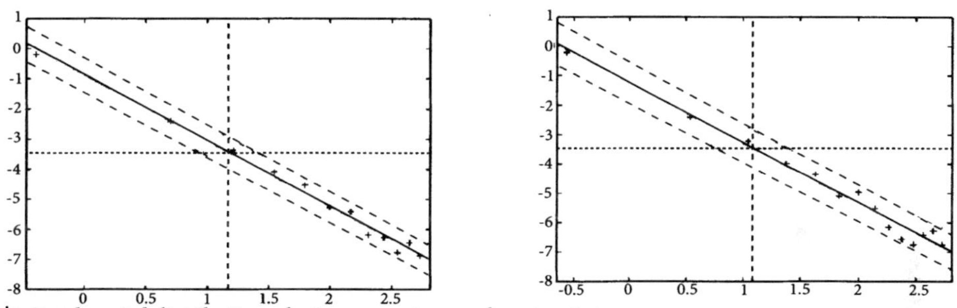

b Total capital distributions during expansions and contractions

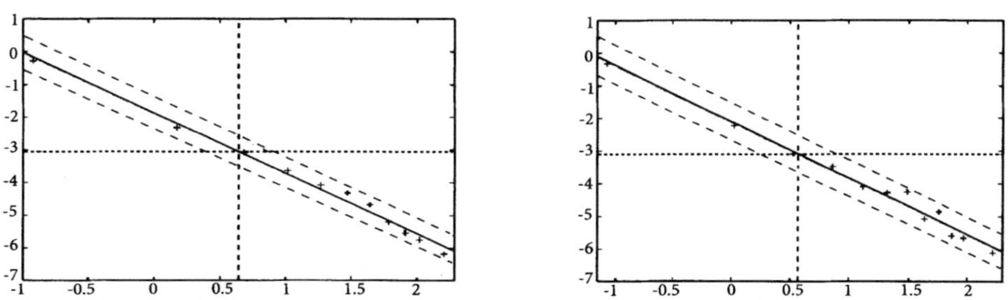

c Total debt distribution during expansions and contractions

Fig. 2.2. Zipf plots of total sales (**a**), of total capital employed (**b**) and of loan capital (**c**), conditioned on expansions (E) and recessions (R)

and for total capital employed during recessions. Furthermore, the scaling exponent for the unconditional distribution seem to be a weighted average of the conditioned scaling exponents in expansions and recessions when firms' size is proxied by total sales and total capital, but not when size is measured by total debt.

A final related feature deserves to be stressed. While the variability in means and standard deviations associated with business cycle fluctuations does not seem to affect the shape of the size distributions, which obey to a Power Law both in expansions and in recessions, the scaling exponents are systematically lower during downturns in comparison to upturns. This means that on average firms are more evenly distributed during expansions than during recessions.

2.5 The Size Distribution Shift over the Business Cycle

The study of the shape and the stability of the size distribution in countries other than the U.S. is here extended by moving from pooled data for very large firms, to samples of medium to large firms at a national level. The source for our data is the Bureau van Dijk's Amadeus commercial dataset, which contains descriptive and balance-sheet data of about 260,000 firms of 45 European countries for the years 1992–2001.

For every firm, juridical, historical and descriptive data are reported (as e.g. year of inclusion, participations, mergers and acquisitions, names of the board directors, news, etc.). Furthermore, Amadeus reports the current values of stocktaking, of balance-sheets (BS), profit and loss accounts (P/L) and financial ratios. The amount and the completeness of available data differ from country to country. To be included in the data set, firms must satisfy at least one of these three-dimensional criteria:

- for UK, France, Germany, Italy, Russian Federation and Ukraine,
  - operating revenue greater or equal to 15 million euro;
  - total assets greater or equal to 30 million euro;
  - number of employees greater or equal to 150;
- for all other countries,
  - operating revenue greater or equal to 10 million euro;
  - total assets greater or equal to 20 million euro;
  - number of employees greater or equal to 100.

The plots reported in Fig. 2.3 are a representative sample of our findings, showing that the size distribution follows a power-law in the range of observation regardless of the proxy we take to measure firms' size. Evidence is reported for the cumulative distributions of total assets (a) and sales (b) in France, and number of employees in UK (c).[8] The power-law fit for $s \geq s_0$,

[8] The number of data points are 8313, 15,776 and 15,055, respectively.

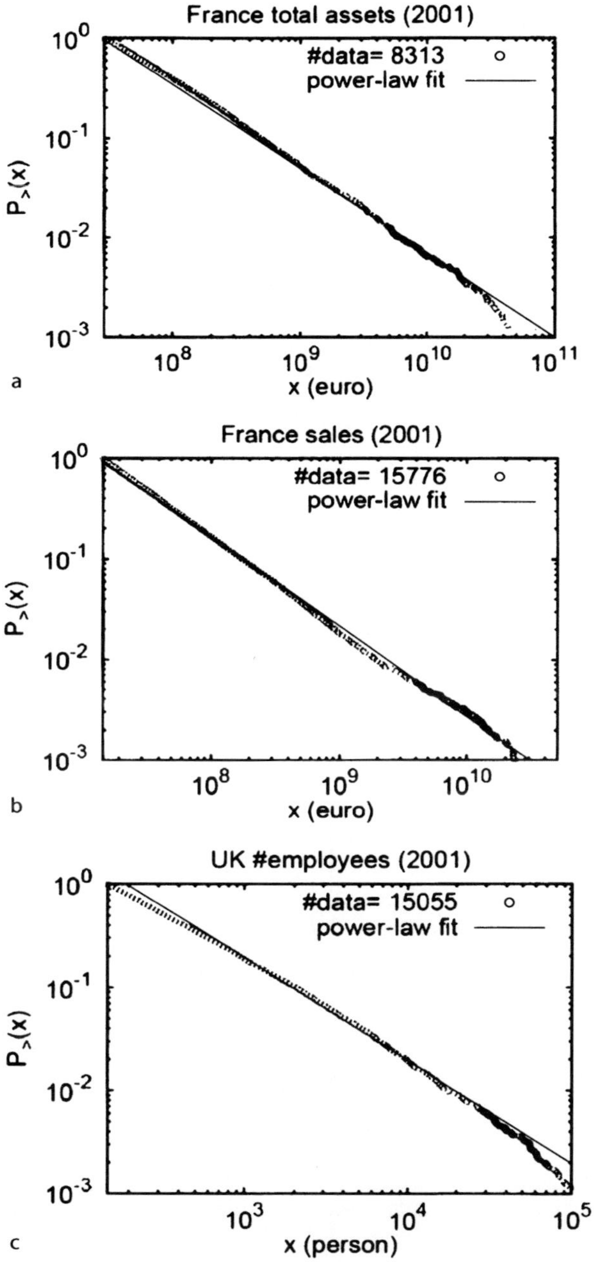

Fig. 2.3. Cumulative distribution of firm's size: **a** France (2001), total assets higher than 30 million euros; **b** France (2001), sales higher than 15 million euros; **c** UK (2001) number of employees in excess of 150 persons

where s_0 denotes the threshold mentioned above, yields the following values of α; (a) 0.886 ± 0.005, (b) 0.896 ± 0.011, (c) 0.995 ± 0.013 (standard error at 99% significance level). The power-law fit is quite good for firms spanning nearly three orders of magnitude.

Figure 2.4 reports the annual change of the size distribution Pareto indices for four countries, namely Italy, Spain, France and UK. The degree of variability of the size distribution over the business cycle seems to be country-dependent.

Figure 2.5 extends the evidence for the country where the variability is higher, that is Italy, by plotting the time series of the (log of the) scale (dashed line) and the shape (continuous line) parameters of the size distribution. Estimates have been obtained by OLS linear fitting on a log-log space. After getting rid of finite sample biases, each regression explains more than 98% of the total variance.

All the series display a significant variability, and changes of the scale and shape parameters are strongly correlated in each case. Notice also that the size distribution measured by different proxies do not tend to move together. The size distribution defined in terms of the number of employees shifts inward and presents a decreasing slope during the recession of the early 1990s,[9] while both the minimum size and the exponent of the power law increase dur-

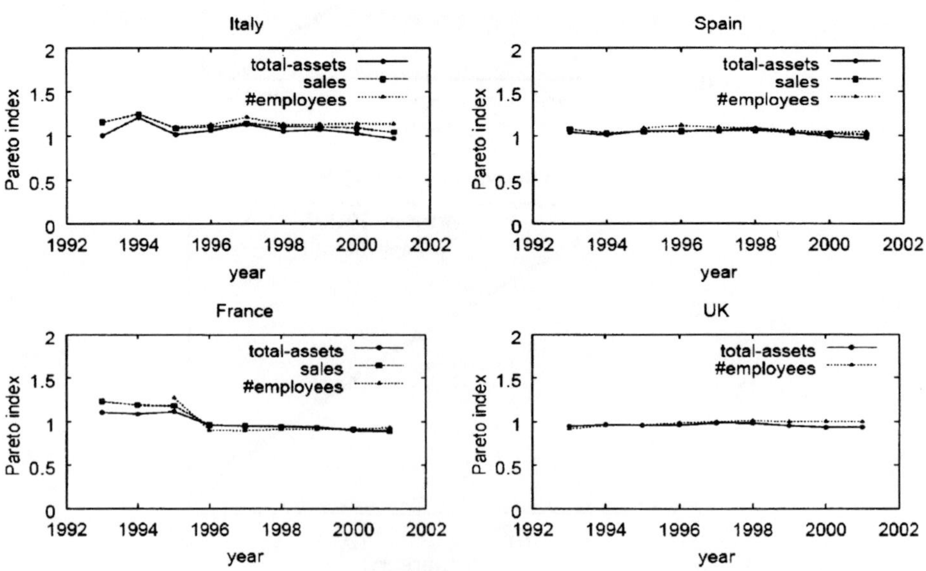

Fig. 2.4. Annual change of Pareto indices for Italy, Spain, France and UK from 1993 to 2001

[9] According to the business cycle chronology calculated by the Economic Cycle Research Institute, Italy experienced a peak in February 1992 and a trough in October 1993. Gallegati and Stanca (1998), using annual data, calculate turning

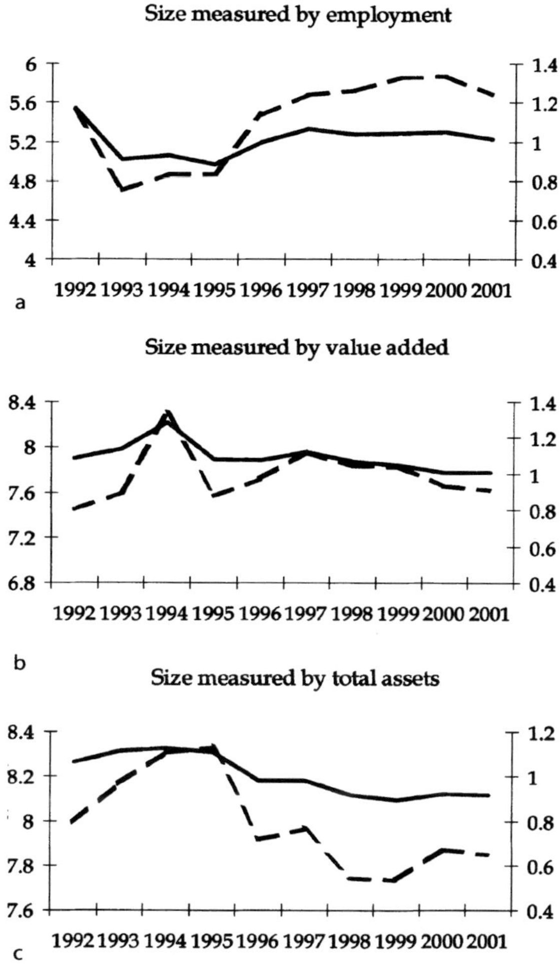

Fig. 2.5a–c. Estimates of the scale parameter (*dashed line, left axis*) and of the shape parameter (*continuous line, right axis*) for the Italian firms, Italy 1992–2001

ing the long expansion of the 1994–2000 period. Movements in the opposite direction are displayed by the size distribution proxied by value added and total assets.

As said before, these results should be interpreted in the light of previous, apparently conflicting, empirical and theoretical work. Amaral *et al.* (1997), for instance, present evidence on the probability density of firms' size as measured by sales for a sample of U.S. firms from 1974 to 1993, showing that the distribution is remarkably stable over the whole period. Matter of factly

points to be in 1990 (peak) and 1993 (trough). It is generally accepted that the following expansion has lengthen at least until the first quarter of 2001.

Amaral and his co-authors recognize that "[...] *there is no existing theoretical reason to expect that the size distribution of firms could remain stable as the economy grows, as the composition of output changes, and as factors that economists would expect to affect firms' size (like computer technology) evolve*" (Amaral *et al.*, 1997, p. 624). Stability of the size distribution, however, is precisely the outcome one should expect according to Axtell (2001). Making use of a random growth process with a lower reflecting barrier studied by Malcai *et al.* (1999), he calculates theoretical power law exponents for the U.S. size distribution measured by the number of employees in each year from 1988 to 1996. It turns out that the hypothesis of a Zipf Law can not be rejected at any standard significance level, the same finding he has obtained empirically for 1997 using more than 5 millions data points from the Census Bureau. It must be incidentally noticed, however, that Axtell's calculations – and therefore his conclusions about the stability over time of the Zipf Law – are biased towards an acceptance of the null of the Zipf Law due to the way the smallest size of the system's components is specified.[10]

Our main points, however, go well beyond this technical drawback. In the following section we shall argue that:

1. Provided that the Pareto distribution represents an attractor for the distribution dynamics regardless of the proxy one uses to measure firms' size, there are indeed theoretical reasons to expect its position and shape to fluctuate over time. Furthermore, even small fluctuations can have important effects;
2. There are also compelling theoretical reasons to expect the fluctuations of the size distribution to diverge as we measure firms' size by recurring to different proxies. Furthermore, such differences represent a key for understanding the nature of the business cycle.

Analytically, let the cumulative distribution of firms' size at time t be given by $F_t(x)$. Time is assumed to be discrete. We can now follow Quah (1993) in associating to each F_t a probability measure λ_t, such that $\lambda_t((-\infty, x]) = F_t(x)$, $\forall x \in \mathbb{R}$. Given that we are working with counter-cumulative distributions, we introduce a complementary measure μ, such that $\mu_t = 1 - \lambda_t = 1 - F_t$. The dynamics of the counter-cumulative size distribution is then given by the stochastic difference equation:

$$\mu_t = V\left(\mu_{t-1}, \varepsilon_t\right) , \tag{2.3}$$

[10] The reason lies in the fact that in Axtell's calculation, the minimum size s_0 has been assumed fixed and equal to 1. From the argument reported in Blank and Solomon (2000), who discuss a paper by Gabaix (1999) who makes use of the same assumption, it emerges that the formula (4) in Axtell (2001) implicitly returns the Pareto exponent α if and only if the minimum size is assumed to be a constant fraction c of the current average of firms' size $\langle s \rangle$, so that one should posit $s_0 = c\langle s \rangle(t)$, which clearly varies in time.

where ε is a disturbance, while the operator V maps the Cartesian product of probability measures with disturbances to probability measures. The empirical evidence discussed above suggests that the invariant size distribution is Pareto so that, for sufficiently large intervals $(s_2 - s_1)$ and h, we impose that:

$$\frac{\sum_{s=s_1}^{s_2} \mu_s}{s_2 - s_1} = \frac{\sum_{s=s_1}^{s_2} \mu_{h+s}}{s_2 - s_1}, \tag{2.4}$$

while at the same time asking whether do there exist theoretical reasons to expect the operator V to fluctuate around its *mean* as business cycle phases alternate.

2.6 Does it Make any Sense?

Despite some work on Pareto distributions' dynamics in the last decade by physicists (see among the others Cont *et al.*, 1997; Joh *et al.*, 1999; Eng *et al.*, 2002; Czirok *et al.*, 1996; Powers, 1998), economists have largely neglected such an issue. Notable exceptions are Brakman *et al.* (1999), who report a n-shape time series for the scaling exponent of Dutch city sizes distribution over more than four centuries, and Mizuno *et al.* (2002), who find that the cumulative distribution of Japanese company's income shifts year by year during the 1970–1999 period, while the scaling exponent of the right tail obeys Zipf Law.[11] In what follows we further elaborate on this issue, with particular regards to fluctuations of the size distribution over the business cycle.

Shifts of the size distribution on a log-log space are related to a change of the system's minimum size (z_0 in (2.1)), which in turn reflects a change of the minimum efficient scale (MES) of operating firms. There are many theoretical reasons to expect the MES to change over the business cycle. Furthermore, changes of the MES over the cycle depends on the proxy we use to measure the operating scale (i.e., the size).

Consider for instance a diffused technical innovation process. While at the aggregate level we observe an increase in total factor productivity, at a microeconomic level we expect a shift of the size distribution by added value, while size distributions by employment and capital should remain stable, or decreasing. In turn, if the technology remains constant in the presence of a product innovation or an increase in demand, than we expect a shit of the value added Pareto distribution, while the size distribution should remain stable if the firms' size is measured by inputs. Finally, if there is a labor saving innovation one expects that the employees distribution shifts towards south-west more than the capital one.

[11] Mizuno *et al.* (2002) prove this last result only by means of visual inspection: no calculations of the scaling exponents are explicitly reported.

According to this approach, one should look at the various distributions not in isolation, but in terms of their relative movements. Meaningfully, relative movements of employment and capital with respect to the value added can be immediately translated into changes of productivity, although these movements should be appropriately disentangled to be fully appreciated. Figure 2.6 reports the relation between labor productivity (roughly measured as the ratio of added value to employment) and firms' size by total assets for Italy between 1996 and 2001, while Fig. 2.7 reports the labor productivity probability density plot on a log-log space.

Three facts clearly emerges from the data. First, there is not clear correlation between labor productivity and firms' size. From the viewpoint of business cycle analysis, the choice of the proxy one uses to measure firms' size is far from neutral. Second, labor productivity seems to be approximated by a Pareto distribution.[12] In other words, labor productivity shares the same distributive features of the size distribution. Third, the distribution of the labor productivity shifts over time. As a first order approximation, a unifying explanation to these facts can be given along Schumpeterian lines. The typical cyclical dynamics should have the following structure: firms follow a directed technical change path by accumulating capital that allows the production of the same output using less quantities of labor as input. The growth of firms' size implied by labor-saving innovations generates a wage increase, due to a positive wage-firm size relationship and, consequently, a shift towards south-west of the firms' size power law distribution in a log-log space. After the wage level has reached its peak, the capital accumulation re-start to grow, while wages diminish and the power law shifts towards north-east.

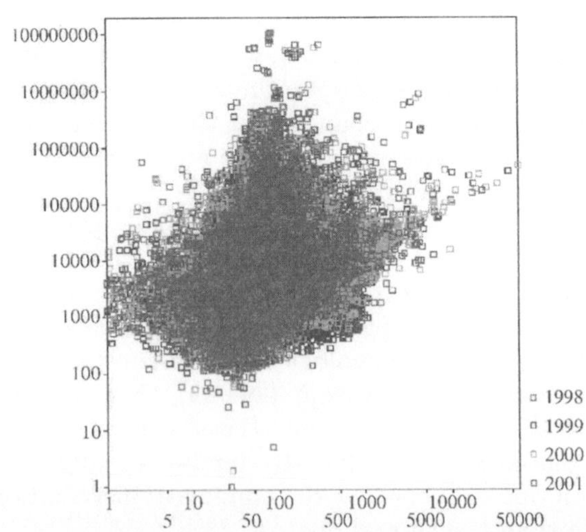

Fig. 2.6. Labor productivity versus corporate firms' size, Italy 1998–2001

[12] More on this in Sect. 2.4.

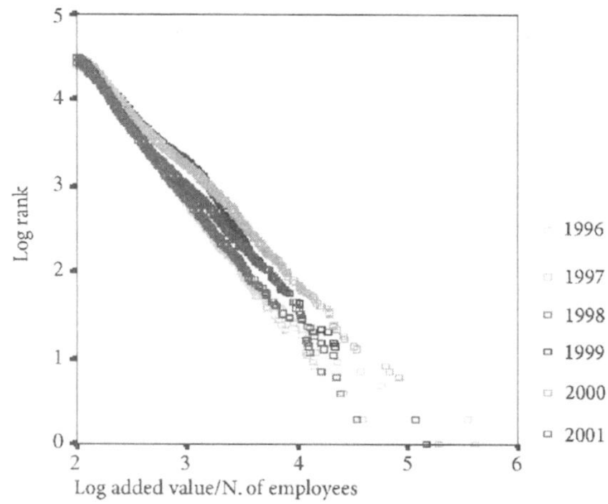

Fig. 2.7. Shift of the distribution of labour productivity for corporate firms. Italy 1996–2001

Size distribution may also shifts because of *firms' demography.* In particular, a major cause of exit is due to bankruptcy, which is likely to affect firms at different scale of operation, as recent examples in U.S. and Italy has taught. Nevertheless, a large amount of empirical evidence has shown that smaller firms are in general more financially fragile (Fazzari *et al.*, 1988).

2.7 Power Laws' Changes in Slope

The evidence reported in Fig. 2.4 highlights that movements of the size distribution over time are not confined to shifts on a log-log plane, but also the slope of the rank-size representation – that is, the scaling exponent of the Pareto-distributed size distribution – fluctuates.

This fact has important implications for a proper understanding of the industry and macroeconomic dynamics from a *structure-conduct-performance* (SCP) perspective. In fact, fluctuations of the scaling exponent of the size distribution immediately translate into fluctuations of the well-known Hirschman–Herfindahl Index (HHI) of industry concentration (Naldi, 2003): the lower the estimated scaling exponent α from the empirical size distribution, the higher the degree of concentration of the supply side of the economy. Under the simplifying assumptions of an economy composed of firms playing a homogeneous Cournot game and of a constant elasticity of demand, fluctuations of the HHI may in turn be associated to fluctuations of the weighted average of the firms' price-cost margins (Cowling and Waterson, 1976), that is fluctuations of markups and profits.

The possibility of slope changes conditioned to business cycle phases for a power law distributed size distribution can be easily proved, depending on the generative process under scrutiny. For instance, let the process generating industry dynamics be given by a simple random multiplicative process:

$$k_i(t+1) = \lambda_i(t)\, k_i(t) \ , \tag{2.5}$$

where $k_i(t)$ is the size of firm i (measured by its capital stock) at time t, and $\lambda_i(t)$ is a random variable with distribution $\Lambda(\lambda, \sigma^2)$. The total number of firms N increases according to a proportionality rule (at each t, the number of new-born firms ΔN, each one with size $k_{\min}$, is proportional to the increase of the economy-wide capital stock K), while firms which shrink below a minimum size (once again $k_{\min}$) go out of business. Blank and Solomon (2000) show that the dynamics of this model converges towards a power law distribution, whose scaling exponent is implicitly defined by the following condition:

$$F = \frac{\Delta K}{k_{\min}\Delta N} = \frac{1}{\left(1 - \frac{1}{\alpha}\right)} \ . \tag{2.6}$$

It seems plausible to expect that the quantity F, which is the inverse of the weight of entrants' contribution to total capital accumulation, changes with the business cycle. In particular, F is likely to increase during recessions (when the number of entrants generally shrinks) and to decrease during expansions. If this assumption is correct, this simple model implies that the scaling exponent of the size distribution fluctuates over the business cycle, to assume lower values during recessions and higher values during expansions, as in real data.

2.8 A Mean/Variance Relationship for the Size Distribution

Another distributional empirical regularity regarding the size distribution relates to the emergence of a scaling relationship between average sizes and cross-sectional volatility of firms, very much in line with a concept – the Taylor's power law (TPL) – firstly associated to biological systems. (Taylor, 1961; Taylor *et al.*, 1978).

The TPL is defined as a species-specific relationship between the temporal or spatial variance of populations $\sigma^2(S)$ and their mean abundance $\langle S\rangle$. Such a relationship turns out to be a power law with scaling exponent β

$$\sigma^2(S) \propto \langle S\rangle^{\beta} \ , \tag{2.7}$$

with (2.7) holding for more than 400 species in taxa ranging from protists to vertebrates over different ecological systems (Taylor and Woiwod, 1982).

The intriguing trait of the TPL does not reside in the scaling relationship *per se*, but in the values assumed by empirical estimates of the scaling exponent β. In fact, from a time series perspective $\sigma^2(S) \propto \langle S\rangle^2$ is precisely what one would expect as soon as populations' dynamics are modelled as homogeneous, independent random processes endowed with finite mean and variance.[13] Thus, an estimated slope lower (higher) than 2 signals that the per capita variability tends to decrease (increase) as the mean population abundance increases. From a spatial perspective, if there exists an equal probability of an organism to occupy a given point in space, populations should be composed of many independent elements leading to a Poisson distribution, which is characterized by a variance-mean ratio equal to 1. It follows that estimates of β higher (lower) than 1 indicate spatial clustering (over-dispersion).

In their seminal contributions, Taylor and his co-authors reported estimates for β for various arthropods ranging from 0.7 to 3.08, but for the majority of species the scaling exponent lies between 1 and 2, a result largely confirmed both in ecological studies (e.g. Anderson *et al.*, 1982; Keitt and Stanley, 1998) and epidemiology (Keeling and Grenfell, 1999). Such an evidence signals that the pattern of spatio-temporal distribution of natural populations is generally characterized by a significant degree of aggregation,[14] but at the same time abundant populations tend to be relatively less variable.[15] Keeling (2000) and Kilpatrick and Ives (2003) provide probabilistic models based on negative interactions among species and spatial heterogeneity aimed at explaining this empirical regularity.

As firms can be plausibly grouped in well defined sectors of activity – or, extending the biological metaphor, *species* – it seems natural to start applying the TPL approach in economics from here. Hence, firms belonging to a certain sector i at year t may be considered as a single population. The relevant characteristic subject to measurement we choose is the members' size, so that we can calculate the mean $\mu_i(t)$ and variance $\sigma_i^2(t)$ of time t firms' size belonging to sector i.

The data we employ have been retrieved from the dataset Amadeus. For the sake of exposition, we select three countries, namely France, Italy and Spain, which could be considered representative of different behaviours in the relevant parameter's space. Firm data cover 18 primary, manufacturing and service industries according to the two digit Nace Rev. 1 classification from year 1996 through 2001.[16] For each country in our sample, we check for

---

[13] Let X be a random variable with finite mean μ and variance σ^2, and k a constant. Then, the mean and the variance of kX are $k\mu$ and $k^2\sigma^2$, respectively. On a log-log plot, the relationship between $k\mu$ and $k^2\sigma^2$ is a line with a slope of 2.

[14] In other words, upon finding one organism/individual there is an increased probability of finding another. In epidemiology, a natural interpretation is given in terms of contagion.

[15] That is, larger populations display a relatively lower probability of extinction.

[16] The nomenclature of the industries (Nace code inside brackets) employed is: 1) Agriculture (A); 2) Manufacture of food products, beverages and tobacco (DA); 3) Manufacture of textiles and textile products (DB); 4) Manufacture of leather and leather products (DC); 5) Manufacture of wood and wood products (DD); 6) Manufacture of pulp, paper and paper products, publishing

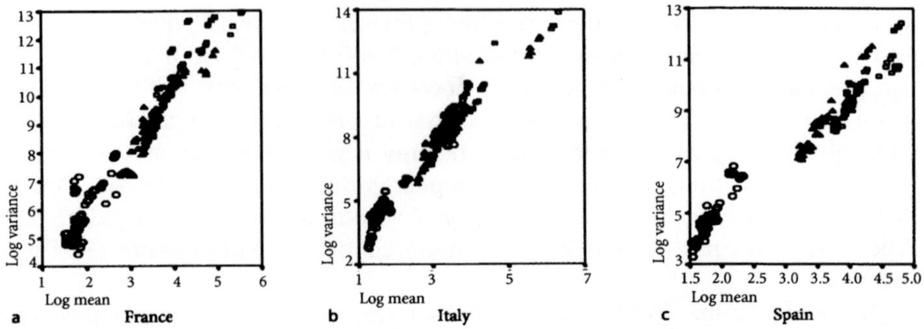

Fig. 2.8a–c. Firms' size variance-mean plots for three European countries. Each point represents the time t ($t = 1996, 2001$) pair log(variance) – log(mean) for firms' size belonging to sector i ($i = 1, \ldots, 18$ as defined in footnote 6), with sizes measured by total assets (*circles*), value added (*squares*) and number of employees (*triangles*), respectively. If the power law (1) holds, data are organized on a linear relationship with positive slope

the existence of a scaling relationship between the mean and the variance of firms' size by considering three alternative measures, i.e. total assets, value added and the number of employees. Hence, for each size measurement we have 108 observations. Results of scatter plots are presented in Fig. 2.8.

From (2.7), it is immediate to note that if the TPL holds the relationship between the log variance and the log mean is linear:

$$\log \sigma^2 = \log a + \beta \log \mu \tag{2.8}$$

with a being a scale parameter. Interestingly enough, for all three countries, and for all the three alternative size measurements as well, a linear relationship emerges neatly. In other terms, besides being typical of natural populations, the TPL seems to characterize the relationship between the mean and the dispersion around it of firms' size.

The linear specification (2.8) implies that its parameters can be consistently estimated by means of OLS. Regression results are reported in Table 2.3. All parameters are statistically significant at the 1% level, and the goodness of fit can be considered largely satisfactory in all cases. With regards to the scaling exponent β, two results deserve to be emphasized. First,

and printing (DE); 7) Manufacture of coke, refined petroleum products and nuclear fuel (DF); 8) Manufacture of chemicals, chemical products and man-made fibres (DG); 9) Manufacture of rubber and plastic products (DH); 10) Manufacture of other non-metallic mineral products (DI); 11) Manufacture of basic metals and fabricated metal products (DJ); 12) Manufacture of machinery and equipment n.e.c. (DK); 13) Manufacture of electrical and optical equipment (DL); 14) Manufacture of transport equipment (DM); 15) Manufacturing n.e.c. (DN); 16) Electricity, gas and water supply (E); 17) Construction (F); 18) Wholesale and retail trade (G).

Table 2.3. OLS estimation results of the TPL parameters, as derived from equation (2.8) in the text. Numbers in parenthesis are standard errors. For each equation, the total number of observations is 108

		Total assets	Value added	# of employees
France	a	1.713 (0.442)	1.200 (0.342)	1.200 (0.342)
	β	2.056 (0.108)	2.161 (0.170)	2.161 (0.170)
	$\bar{R}^2$	0.903	0.815	0.815
Italy	a	−3.177 (0.562)	−3.427 (0.475)	−2.095 (0.287)
	β	3.089 (0.135)	3.326 (0.132)	3.822 (0.159)
	$\bar{R}^2$	0.929	0.941	0.936
Spain	a	1.191 (0.261)	1.820 (0.27)	1.327 (0.151)
	β	1.905 (0.068)	1.905 (0.067)	1.940 (0.084)
	$\bar{R}^2$	0.952	0.953	0.931

for each country size measurements are quantitatively equivalent. Second, the slope of the TPL in its log-linear version differs substantially across countries. The estimated β turns out to be slightly below 2 for Spain, somewhat higher than 2 for France, and well above 3 for Italy.

3 Stylized Facts in Industrial Dynamics: Exit, Productivity, Income

3.1 The Exit of Firms

In principle, a firm can go out of business as an independent unit for three reasons: i) voluntary exit, due for example to the prospective of unsustainable reduction of profitability; ii) merger with another firm or acquisition by another firm; iii) bankruptcy due to the inability of a firm to pay its financial obligations as they mature.

Empirical studies on the determinants of firms' exit have long noted that the probability of survival to events i) and iii) appears to increase with age and size, and that industry characteristics do not affect significantly the probability of survival (Siegfried and Evans, 1994; Caves, 1998). These results, which contrast with the prediction of the purely random-driven Gibrat's law of proportional effect, are in fact fully consistent with existing behavioral accounts of the firm's life cycle, including both passive learning models (Jovanovic, 1982; Hopenhayn, 1992) and active learning (Ericson and Pakes, 1995) models.

The strong positive dependence of the probability of survival on size reported in the empirical literature on firm dynamics leads naturally to ask whether very large corporations may actually fail. For instance, Marshall (1920) seems dubious on the eventual demise of large firms: "[...] *vast joint stock companies* [...] *often stagnate but do not readily die*". However, a cursory look at the available evidence seems to return a picture which is at odds with Marshall's view. Hannah (1999), for example, constructs a data set listing the 100 largest industrial companies in the world in 1912. By 1995, only 52 of these firms survived in any independent form. Furthermore, 24 out of the 52 survivors were smaller than they were in 1912. All in all, very large firms do in fact die. The search for additional evidence in different datasets on the exit of firms may therefore be useful for addressing this and other interesting issues. This is precisely what we do in the following.

3.1.1 Evidence on the Extinction Rate

Scaling plot techniques have been recently applied to research on firms' extinction rate. The main example is the paper by Cook and Ormerod (2003) (henceforth, CO), who present evidence of power law scaling for the demises of US firms. In particular, CO show that the exit rate approximately follows a power law distribution with exponent close to 2 by sector. Interestingly enough, this value is very much in line with the literature on the Raup-Sepkoski's kill curve, according to which biological extinction events "[...] *can be reasonably well fitted to a power law with exponent between 1 and 3*" (Bak, 1997, p. 165). In this Section we apply the same methodology to data retrieved from the OECD firm-level data project, regarding demises of firms in eight OECD countries in the period 1977–1999.[1]

The dataset contains information on the frequency of firm demises on an annual basis, split into 40 different industrial sectors for each of the eight countries (see Bartelsman *et al.*, 2003). Demises are then expressed in terms of frequencies, as we divide the total number of exits by the total number of operating firms. This gives rise to a total number of 5051 observations. We call each of these observations a *group*: each group specifies a particular industry in a particular year in a particular country.

The correlation of extinction rates in the same sector across countries is quite high (0.6), while the temporal correlation and the correlations across sectors and countries are much lower (0.19 and 0.15, respectively). These figures suggest that the exit of firms is probably driven by sectoral shocks instead of country-specific shocks, and that the overall state of the economy plays a minor role in determining demises.

It must be noted that, while the data we employ have annual frequency, exits during a year occur on a daily basis. The statistical model we adopt as a benchmark for empirical analysis postulates that each observation of the sample is treated as the sum of 250 independent and identically distributed random variables (250 being the approximate number of working days per year). If our assumption is correct, according to the law of large numbers the distribution of firms' exit will be Gaussian. As shown in Fig. 3.1, however, if any convergence to a Gaussian distribution occurs it seems to be extremely slow. More formal statistical analysis supports the conclusion one can attain by visual inspection. In fact, a Kolmogorov–Smirnov test rejects the null hypothesis of normality at the 1% significance level.

Due to the significant kurtosis and the high weight on the right tail showed by the empirical distribution, as an alternative to the Gaussian we propose a truncated power law model with exponent r. To test the hypothesis of

[1] The countries in our sample are Denmark, Finland, France, Italy, Netherlands, Portugal, United Kingdom and United States. The data set covers varying time spans over the period 1980–1998, even if it mainly refers to the period 1989–1994. Firm is individuated, adopting the Eurostat definition (Council Regulation (EEC) No 696/93), as "[...] *an organizational unit producing goods or services which benefits from a certain degree of autonomy in decision-making, especially for the allocation of its current resources*". All single-person business were not considered. The industry classification follows the OECD Structural Analysis Database (STAN).

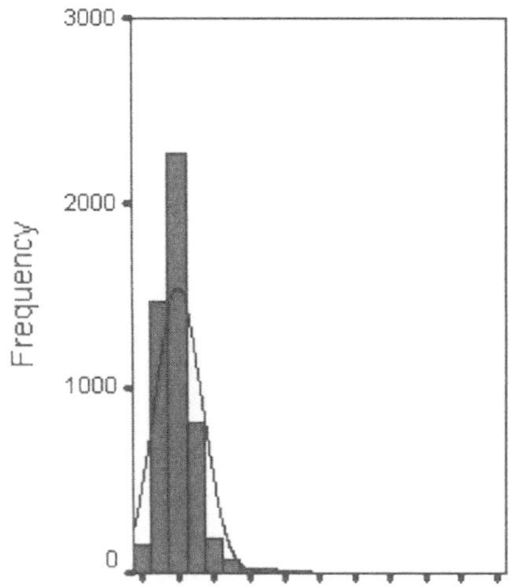

Fig. 3.1. Absolute frequency of extinctions on total number of firms (mean = 0.1, s.d. = 0.07)

a power law distribution for daily occurrences, for each annual observation we register the occurrence of a random variable generated by the following probability distribution:

$$P(X = i) = \frac{(i+1)^{-r}}{\sum_{j=0}^{m} (j+1)^{-r}} \tag{3.1}$$

with $i, j = 0, \ldots, m$, where m is the maximum of daily demises. Notice that a finite value for the parameter m allows us to set the variance of the distribution to a finite value.

After obtaining a sample of 5051 simulated data-points, we plot them against the actual data. We perform alternative simulations for the two parameters r and m varying on a wide range. The highest correlation between simulated and actual data has been obtained for $r = 2$ and $m = 400$, with a correlation coefficient equal to 0.992 (Fig. 3.2). The Kolmogorov–Smirnov test confirms that the two samples come from the same distribution at the standard significance level. Although the central part of the theoretical dis-

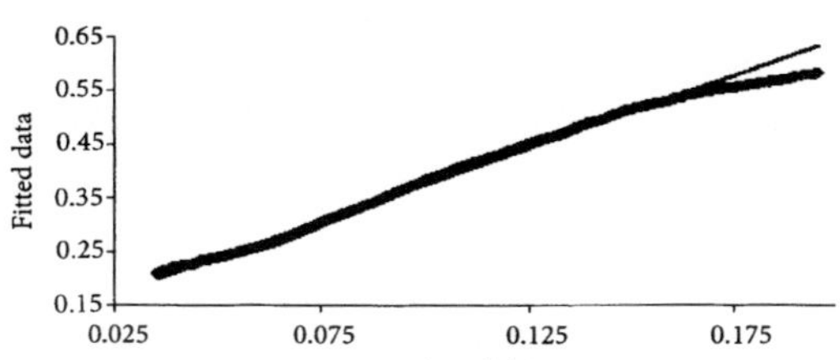

Fig. 3.2. Scatter plot of real data and data drawn from a truncated power law distribution

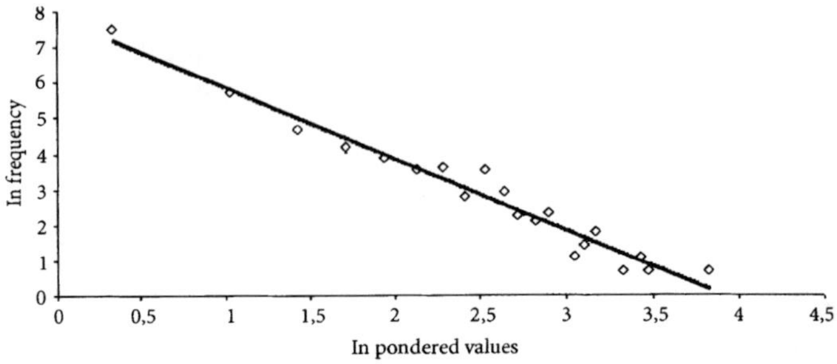

Fig. 3.3. Zipf plot of pondered values of annual demises

tribution displays a remarkable good fit to the actual one, the tails tend somehow to overestimate the upper part and to underestimate the lower part of the distribution.

As a further test to check the robustness of the our results, in Fig. 3.3 we report the Zipf plot of the absolute values of annual demises, weighted with their relative frequency, on their absolute frequency. Also in this case, the interpolation line, which returns a R^2 equal to 0.962, implies a decay coefficient very close to two (2.003).

As an additional issue, we notice that exits due to bankruptcy and voluntary shut down are a major determinant of output and employment contraction or, in other terms, are likely to be somehow related to aggregate downturns. Note, in particular, that changes in the aggregate demise rate are not characterized by strong auto-correlation, while disaggregated data show a strong temporal dependence among the same size class. Thus, we test the hypothesis that demises of firms grouped by size are fitted by a Weibull distribution, that is the same distribution which seems to characterize the magnitude of business cycle phases.

The data set we examine reports the demises sorted by number of employees[2] for a total of 548 observations for 9 countries.[3] Again, each observation represents a group, identified by year, country and size. Groups are then divided into five classes sorted by firm sizes.

For any class, the Kolmogorov–Smirnov test rejects the hypothesis that the sample data are normally distributed at standard significance level. Even restricting the sample to a merely 5 percent around the mean, results do not change. Thus, as an alternative we consider a Weibull distribution for the five classes of demises by sizes (in their absolute values). The Weibull cumulative

[2] The firms are sorted in six classes by number of employees (less than 20, from 20 to 49, from 50 to 99, from 100 to 499, more than 500).

[3] In this section we extend the analysis to also West Germany, which was not included in the previous section because of the absence of data relative to the first dimensional class that may cause distortions in the aggregate (and thus in the percentage) of demises in that country.

distribution function takes the form:

$$F(x) \equiv 1 - \exp\left(-\alpha x^{\beta}\right) , \tag{3.2}$$

where α is the scale parameter and β is the characteristic shape parameter that quantifies the speed of decay of the distribution (Malavergne *et al.*, 2003). Notice that if we rank n observations from the largest to the smallest and we indicate with x_i the ith observation ($x_1 > x_2 > \ldots > x_n$), we obtain:

$$\frac{i}{n} \equiv 1 - F(x_i) . \tag{3.3}$$

Substituting (3.2) in (3.3) and taking the natural logarithm yields:

$$x^{\beta} \equiv -\frac{1}{\alpha} \ln(i) + \frac{\ln(n)}{\alpha} . \tag{3.4}$$

Finally, setting $\frac{\ln(n)}{\alpha} \equiv \varphi$ and $\frac{1}{\alpha} \equiv \lambda$, we get:

$$x^{\beta} \equiv -\lambda \ln(i) + \varphi , \tag{3.5}$$

that is the interpolation line on the semi-log plane, as we plot the variable x taken at exponent β against the natural logarithm of its rank. We may also identify an additional parameter

$$x_0 = \left(\frac{1}{\alpha}\right)^{\frac{1}{\beta}}$$

that represents a reference scale, from which all the moments can be determined.

Estimates by OLS of (3.5) on the semi-log plane return a good fit for each class (R^2 is more than 0.96 in every case), although the right tails are systematically overestimated, a result which may be due to finite size effects. Table 3.1 reports the results of the estimation.

Recall that if $\beta = 1$, the Weibull distribution coincides with the exponential. As β goes below 1 the distribution becomes flat while, for $\beta > 1$, observations are concentrated around their mode and the distribution may

Table 3.1. Parameter estimates for the Weibull model of firm demises by size

Classes	x_0	β	λ
> 0	7,0448	1,1469	4,9627
> 20	6,0498	1,2989	3,5243
> 50	4,36	1,164	3,0426
> 100	3,0555	0,9978	2,5552
> 500	2,022	0,787	1,9492

resemble the Gaussian. The trend of the shape parameter β as we move from small to large firms shows that the tail becomes flatter, a result confirmed also by the slope of the regression line.

Another way to look at our results is in terms of the hazard function, that is the "memory" or persistence of the distribution (see Lancaster, 1992), that becomes null (the so-called *lack of memory* property) as $\beta = 1$, and positive (negative) if β greater (smaller) than 1. In our case, our parameter estimates suggest that the probability to record a number $x + \xi$ of demises, once a number x has been already recorded, progressively diminishes as we focus on greater firms.

3.1.2 Power Law for Bad Debt

In spite of the strong evidence supporting the hypothesis that bankruptcies are negatively correlated with size, large firms are far from immune from default. The recent examples of Enron and Worldcom corroborate this claim. In fact, the available evidence (Platt, 1985) is clear-cut in suggesting that

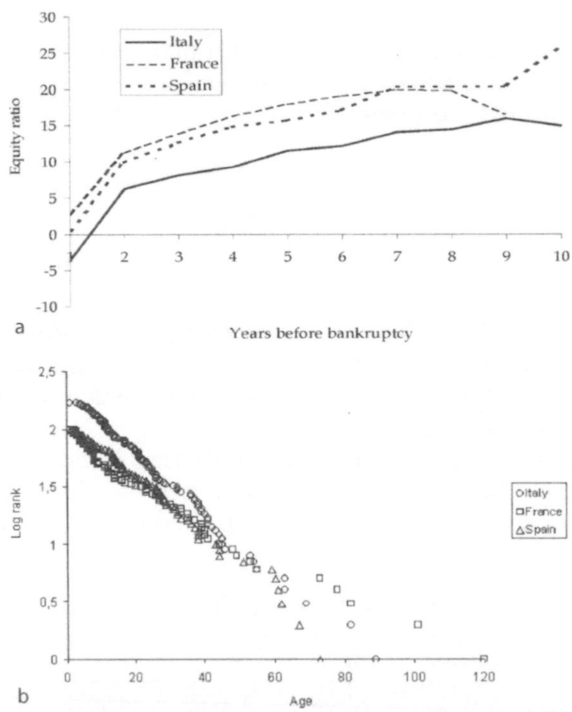

Fig. 3.4. a Profile analysis of the average equity base of bankrupt firms in each of the year before bankruptcy. Data points are referred to mean values for 676 Italian, 1786 French and 750 Spanish firms went bankrupt during the 1992–2001 period. **b** Semi-log plot of the distribution of bankrupt firms by age

insolvencies occur at all scales, and that the proportion of failures varies significantly over time.

To shed some light on the issue of the vulnerability of large companies to bankruptcy, we analyze the available evidence for bankruptcies in a sample of European countries, namely Italy, Spain and France. Data are retrieved from the dataset AMADEUS from 1992 through 2001.

First, we find that financial ratios are invariably a good predictor of firms failure, and therefore exit. In particular, the equity ratio, defined as the ratio of net worth (current assets minus current liabilities) to total assets, deteriorates almost monotonically as the date of bankruptcy approaches (Fig. 3.4a). The distribution of exits by age turns out to be exponential, signaling that the probability to fail is independent of time (Fig. 3.4b). Given that firms generally enter at a small scale, and that they grow over time through investments, this suggests that big and small firms should have a rather equal probability to go out of business.

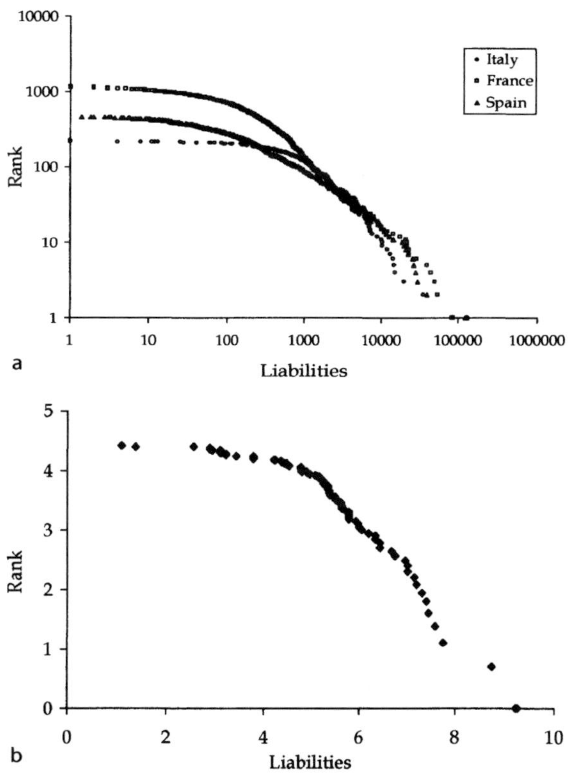

Fig. 3.5. a Zipf plot for the debt of bankrupt firms in Italy (217 firms), France (1166), and Spain (455) during 1999. **b** Zipf plot for European corporate long term debt defaults, from January 1985 through May 2002

As far as debt is concerned, it is interesting to note that the right tail of the distribution of debt of bankrupt firms scales down as a power law for all sampled countries (Fig. 3.5a), $Q(B > b) \sim b^{-\alpha}$. In particular, the scaling exponents for the 60% right tails are $\alpha = -1.09$ for Italy, $\alpha = -0.87$ for France and $\alpha = -0.67$ for Spain. Furthermore, a quantitatively similar scaling exponent results also for data on defaulted debt collected from a different dataset, that is for the European corporate long term debt defaults occurred from January 1985 through May 2002, as reported by Moody's (Fig. 3.5b). The bad debt of insolvent bond issuers is distributed as a power law with $\alpha = -0.92$. All in all, the power law seems to be an invariant structural pattern for the bad debt of European companies.

Indeed, our findings are strikingly close to the ones reported in Aoyama *et al.* (2000) and Fujiwara et al. (2003) for Japanese bankrupt firms, with the bad debt for large failed firms (i.e. the right tail of the distribution) being estimated to scale down with an exponent α comprised between 0.91 and 1. This result strongly suggests that universality, as defined in statistical physics, seems to hold for the bad debt distribution. Any reasonable model of industrial dynamics should take this evidence into account.

3.2 Productivity and Income

The research agenda which drives our empirical exploration consists in thinking about macroeconomic interconnections in terms of distribution functions and their dynamics. Paraphrasing Steindl (1965), a concrete example of such an approach deals with the relationships among the distributions of productivity and personal income. In a nutshell, the productivity of a firm is the key to its average profitability. Profits, in turn, are distributed among firms' owners contributing to their personal income. This information, combined with the positive relationship displayed by data between firm' size and the wage paid to employees on the one hand, and what we know about the size distribution of firms on the other one, determines the distribution of households' income.

3.2.1 The Distribution of Productivity in France and Spain

The recent availability of huge sets of longitudinal firm-level data has generated a number of studies on productivity.[4] In this section we consider two basic measures of productivity: labour and capital productivity. The productivity of labour is defined as the ratio of value added to the number of employees (where value added, defined according to standard balance sheet reporting, is the difference between total revenue and the cost of inputs excluding the cost of labour). Although elementary, this measure has the advantage of being accurately approximated from the available data.

[4] See *inter alia* Hulten et al. (2001) and Kruger (2003).

The productivity of capital is defined as the ratio of value added to the value of fixed assets (i.e. the capital stock). This measure of productivity has some drawbacks since the value of firms' assets changes continuously in time. The volatility of the Stock price, for instance, affects the market value of the capital stock. Usually the industrial organization literature recognizes that the productivity distribution is not normally distributed, and empirical fat tails with power law behaviour are observed.

Figures 3.6 to 3.9 show the log-log plot of the frequency distributions (left) and the complementary cumulative distributions (right) of labour productivity and capital productivity measured as ratios of total added value of the firms. In these figures the different data sets correspond to the years 1996–2001 for two different countries: France and Italy. The frequency distributions show a very clear non-Gaussian character: they are skewed with asymmetric tails and the productivity (Figs. 3.6–3.9 (left)) exhibits a leptokurtic peak around the mode with 'fat tails' (for large productivities) which show a rather linear behaviour in a log-log scale. In these figures we also report, for comparison, the linear trend corresponding to power-law distributions ($N(x) \propto x^{-\alpha}$ with $\alpha = 2.7, 2.1, 3.8$ and 4.6, respectively).

The complementary cumulative distributions ($P_>(x)$, being the probability to find a firm with productivity larger than x) also show a linear trend at large x (in log-log scale) implying a non-Gaussian character with the probability for large productivities well mimicked by a power-law behaviour. The 'fat tails' character of such distributions is highlighted in the inserts of Figs. 3.6–3.9 (right) where log-normal plots show that the decay of $P_>(x)$ with x is much slower than exponential.

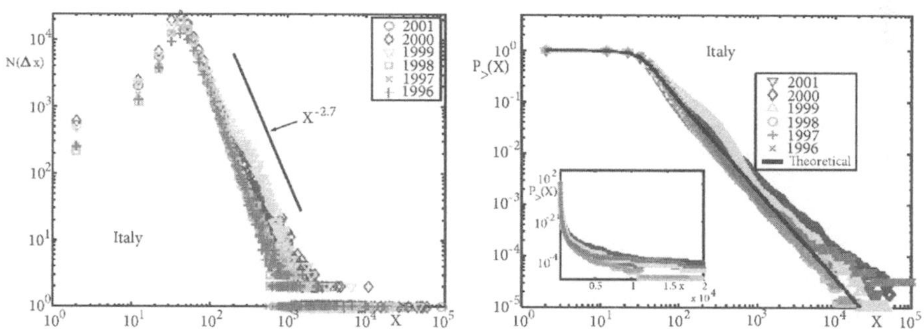

Fig. 3.6. Frequency distributions (*left*) and complementary cumulative distributions (*right*) for the labour productivity in Italy in the years 1996–2001. The theoretical behavior is for $\alpha = 2.7$, $m = 22$, $n = 11$, $\sigma = 10$ and $\beta = 3$. The *insert* shows $P_>(x)$ v.s. x in log-normal scale

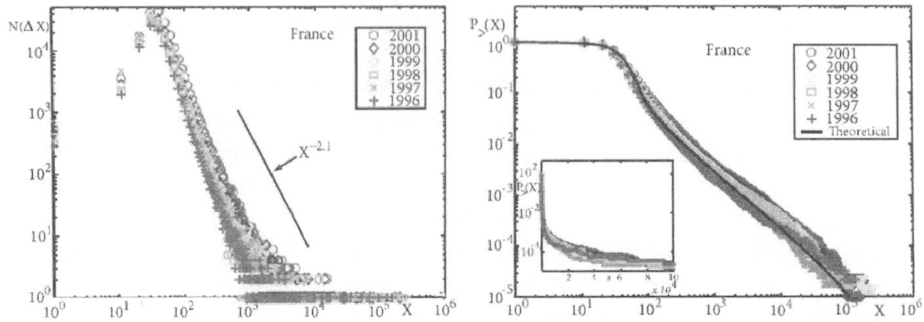

Fig. 3.7. Frequency distributions (*left*) and complementary cumulative distributions (*right*) for the labour productivity in France in the years 1996–2001. The theoretical behavior is for $\alpha = 2.1$, $m = 30$, $n = 4$, $\sigma = 20$ and $\beta = 1$. The *insert* shows $P_>(x)$ v.s. x in log-normal scale

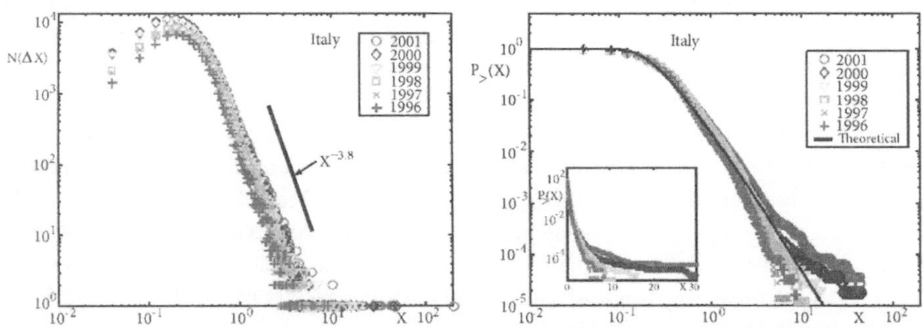

Fig. 3.8. Frequency distributions (*left*) and complementary cumulative distributions (*right*) for the capital productivity in Italy in the years 1996–2001. The theoretical behaviour is for $\beta = 3.8$, $m = 0.04$, $n = 0.02$, $\sigma = 0.01$ and $\beta = 25$. The *insert* shows $P_>(x)$ v.s. x in log-normal scale

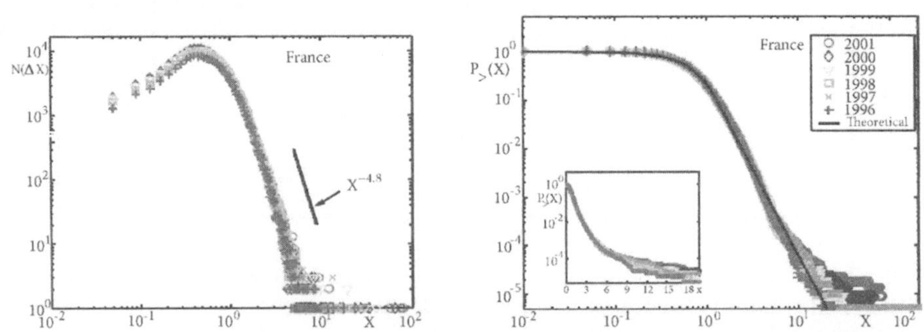

Fig. 3.9. Frequency distributions (*left*) and complementary cumulative distributions (*right*) for the capital productivity in France in the years 1996–2001. The theoretical behaviour is for $\alpha = 4.6$, $m = 0.06$, $n = 0.02$, $\sigma = 0.4$ and $\beta = 68$. The *insert* shows $P_>(x)$ v.s. x in log-normal scale

3.2.2 Power Law Tails in the Italian Personal Income Distribution

The first appearance of the power law distribution in economics is due Vilfredo Pareto, who observed in his *Course d'économie politique* (1897) that a plot of the logarithm of the number of income-receiving units above a certain threshold against the logarithm of the income yields points close to a straight line. Recent empirical work seems to confirm the validity of Pareto (power) law. For example, Aoyama *et al.* (2000) show that the distribution of income and income tax of individuals in Japan for the year 1998 is very well fitted by a Pareto power-law type distribution, even if it gradually deviates as the income approaches lower ranges.

The applicability of Pareto distribution only to high incomes is actually acknowledged; therefore, other kinds of distributions have been proposed by researchers for the low-middle income range. According to Montroll and Shlesinger (1983), US personal income data for the years 1935–1936 suggest a power-law distribution for the high-income range and a lognormal distribution for the rest; a similar shape is found by Souma (2001) investigating the Japanese income and income tax data for the high-income range over the 112 years 1887–1998, and for the middle-income range over the 44 years 1955–1998. Nirei and Souma (2004) confirm the power-law decay for top taxpayers in the US and Japan from 1960 to 1999, but find that the middle portion of the income distribution has rather an exponential form; the same is proposed by Drăgulescu and Yakovenko (2001) for the UK during the period 1994–1999 and for the US in 1998.

In this section we look at the shape of the personal income distribution in Italy by using cross-sectional data samples from the population of Italian Households during the years 1977–2002. We use microdata from the Historical Archive (HA) of the Survey on Household Income and Wealth (SHIW) made publicly available by the Bank of Italy. All amounts are expressed in thousands of Italian Lire. Since we are comparing incomes across years, to get rid of inflation data are reported in 1976 prices using the Consumer Price Index (CPI) issued by the National Institute of Statistics. The average number of income-earners surveyed from the SHIW-HA is about 10,000.

Figure 3.10 shows the profile of the personal income distribution for the year 1998. On the horizontal axis we report the logarithm of income in thousands of Lire and on the vertical axis the logarithm of the cumulative probability.

Two facts emerge from this figure. First, the central body of the distribution – matter of factly almost all of it, i.e. the distribution up to the 99th percentile – follows a two-parameter lognormal distribution, whose probability density function is:

$$f(x\,|\mu,\sigma) = \frac{1}{x\sigma\sqrt{2\pi}} \exp\left\{-\frac{1}{2}\left[\frac{\ln(x)-\mu}{\sigma}\right]^2\right\} \tag{3.6}$$

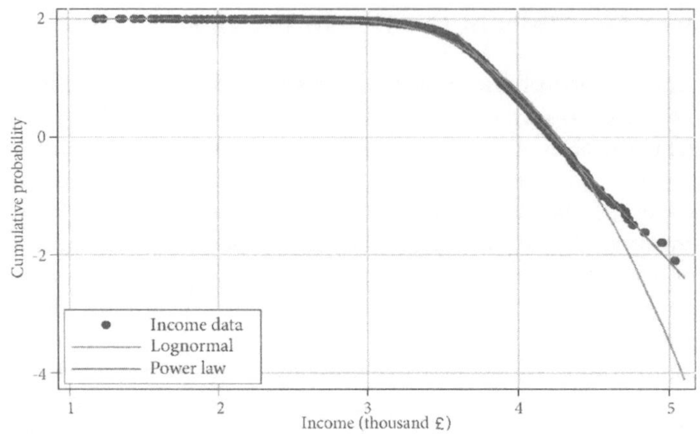

Fig. 3.10. Cumulative probability distribution of the Italian personal income in 1998

with $0 < x < \infty$, μ and σ^2 are the first two moments of the distribution. The value of the fraction $\beta = 1/\sqrt{2\sigma^2}$ returns the so-called Gibrat index; if β is relatively low (i.e. if the variance is high), the personal income is unevenly distributed. From our dataset we obtain the following maximum-likelihood estimates:[5] $\hat{\mu} = 3.48$ (0.004) and $\hat{\sigma} = 0.34$ (0.006);[6] the Gibrat index is $\hat{\beta} = 2.10$.

Second, about the top 1% of the distribution follows a Pareto distribution. This power law behaviour of the tail of the distribution is evident from Fig. 3.11, where the red solid line is the best-fit interpolating function. We extract the power law slope by running a simple OLS regression, obtaining a point estimate of $\hat{s} = 1 + \hat{\alpha} = 2.76$ (0.002). Given this value for s, our estimate of x_0 (the income-level below which the Pareto distribution would not apply) is 17,141 thousand Lire. The fit of the linear regression is extremely good, as one can appreciate by noting that the value of the R^2 index is 0.99.

The pattern of the distribution of personal income consisting of a lognormal for most of the distribution and a power law for the tail seems to hold over the entire time span, as one can easily realize from Fig. 3.12, which reports the shape of the income distribution for all the years. The corresponding estimated parameters for the lognormal and Pareto distributions are given in Table 3.2. The table also shows the values of the Gibrat index. Note, however, that the scaling exponent of the power law and the curvature of the lognormal fit change from year to year.

---

5 We excluded from the sample used for estimation the top 1.4% of the distribution, which we considered an outlier, and about the bottom 0.8%, which corresponds to non-positive entries.

6 The number in parentheses represents the standard error.

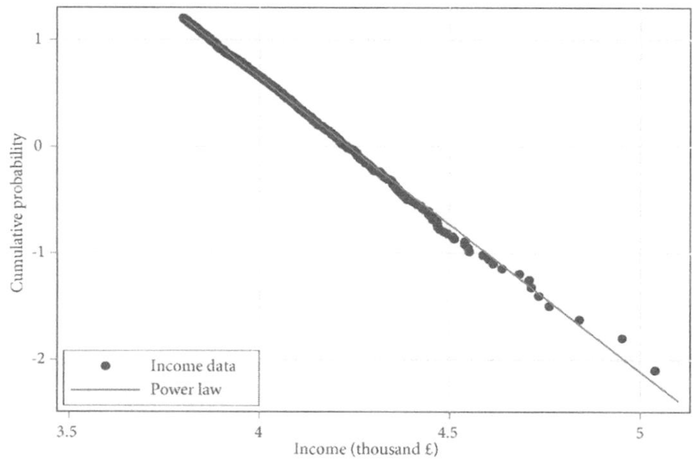

Fig. 3.11. The fit to the power distribution for the year 1998

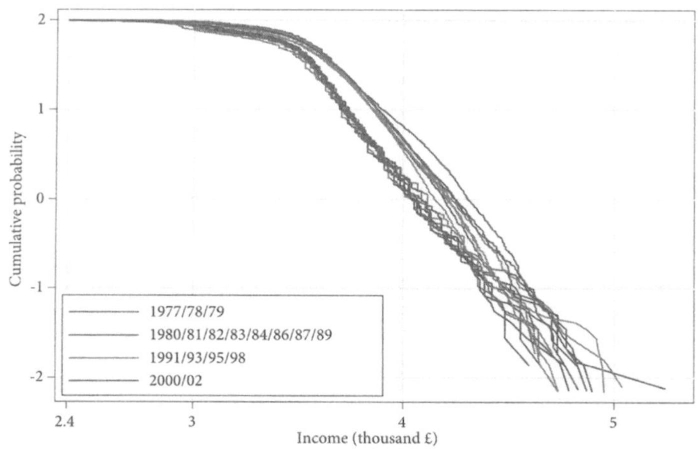

Fig. 3.12. Time development of the Italian personal income distribution over the years 1977–2002

In order to quantify the fluctuations of income distribution from year to year, we start by noticing that the origin of these fluctuations are probably related to changes in the growth rate of the gross domestic product (GDP). To evaluate this conjecture, we study the fluctuations in the growth rates of GDP and personal incomes (PI), and try to show that similar mechanisms may be responsible for the observed growth dynamics of income for both the aggregate economy and individuals. The distribution of the GDP annual growth rates is shown in Fig. 3.13. The data are retrieved from the OECD Statistical Compendium. The growth rates are defined as log-differences, $R_{\mathrm{GDP}} = \log(\mathrm{GDP}_{t+1}/\mathrm{GDP}_t)$. Data are reported in 1976 prices; moreover,

Table 3.2. Estimated lognormal and Pareto distribution parameters for all the years

Year	$\hat{\mu}$	$\hat{\sigma}$	$\hat{\beta}$	$\hat{s}$	$\hat{x}_0$	R^2
1977	3.31 (0.005)	0.34 (0.004)	2.08	3.00 (0.008)	10,876	0.9921
1978	3.33 (0.005)	0.34 (0.004)	2.09	3.01 (0.008)	11,217	0.9933
1979	3.34 (0.005)	0.34 (0.005)	2.08	2.91 (0.009)	11,740	0.9908
1980	3.36 (0.005)	0.33 (0.005)	2.15	3.06 (0.008)	11,453	0.9915
1981	3.36 (0.005)	0.32 (0.004)	2.23	3.30 (0.008)	10,284	0.9939
1982	3.38 (0.004)	0.31 (0.005)	2.27	3.08 (0.005)	11,456	0.9952
1983	3.38 (0.004)	0.30 (0.004)	2.32	3.11 (0.006)	11,147	0.9945
1984	3.39 (0.004)	0.32 (0.005)	2.24	3.05 (0.007)	11,596	0.9937
1986	3.40 (0.004)	0.29 (0.006)	2.40	3.04 (0.005)	11,597	0.9950
1987	3.49 (0.004)	0.30 (0.004)	2.38	2.09 (0.002)	24,120	0.9993
1989	3.53 (0.003)	0.26 (0.003)	2.70	2.91 (0.002)	15,788	0.9995
1991	3.52 (0.004)	0.27 (0.004)	2.58	3.45 (0.008)	14,281	0.9988
1993	3.47 (0.004)	0.33 (0.004)	2.15	2.74 (0.002)	16,625	0.9997
1995	3.46 (0.004)	0.32 (0.003)	2.19	2.72 (0.002)	16,587	0.9996
1998	3.48 (0.004)	0.34 (0.006)	2.10	2.76 (0.002)	17,141	0.9993
2000	3.50 (0.004)	0.32 (0.004)	2.20	2.76 (0.002)	17,470	0.9994
2002	3.52 (0.004)	0.31 (0.005)	2.25	2.71 (0.002)	17,664	0.9997

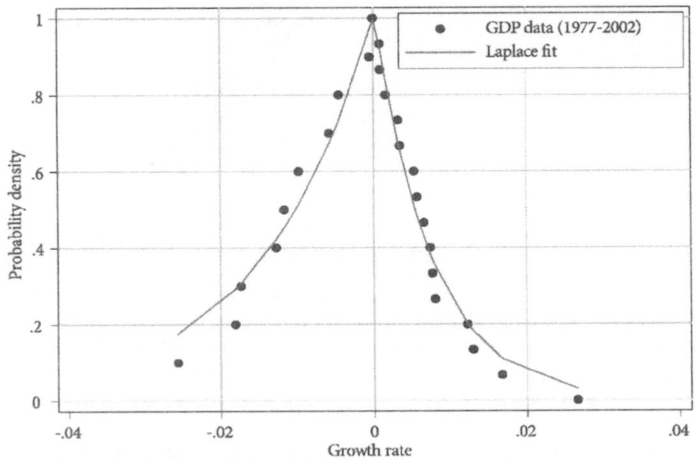

Fig. 3.13. Probability density function of Italian GDP annual growth rates, 1977–2002, together with the Laplace fit (*solid line*)

to improve comparison of the values over the years we detrend them by applying the Hodrick–Prescott filter.

By means of a nonlinear algorithm, we find that the probability density function of annual growth rates is well fitted by a Laplace distribution. This result seems in agreement with the growth dynamics of PI, as shown in

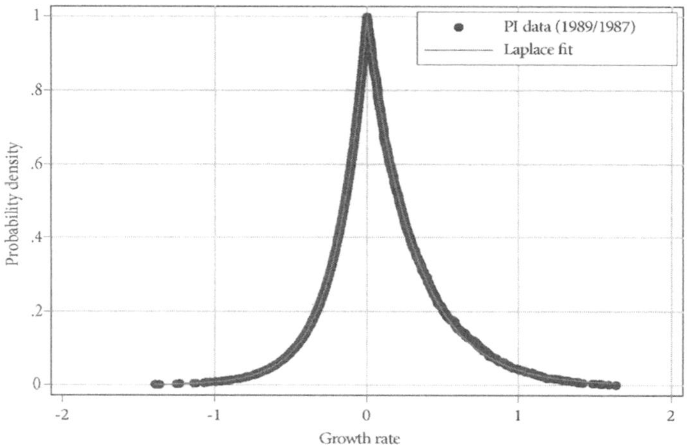

Fig. 3.14. Probability distributions of the Italian PI for the years 1987/1989 ans 1991/1993

Fig. 3.14 for a randomly selected yearly sample. These finding leads us to conjecture that the two phenomena obey the same distribution. Before testing this conjecture, in order to consider almost the same number of data points for the two datasets we draw a 2% random sample of the data we have for individuals, and normalize it using the transformations $(R_{\rm PI} - \bar{R}_{\rm PI})/\sigma_{\rm PI}$ and $(R_{\rm GDP} - \bar{R}_{\rm GDP})/\sigma_{\rm GDP}$. As shown in Table 3.3, which reports the p-values for all the cases we studied, the null hypothesis of equality of the two distributions cannot be rejected at the usual 5% marginal significance level. Therefore, the data are consistent with the assumption that a common empirical law might describe the growth dynamics of both GDP and PI, as shown in Fig. 3.15, where all curves for both GDP and PI growth rate normalized data almost collapse onto the red solid line representing the non-linear Laplace fit.[7]

Table 3.3. Kolmogorov–Smirnov test p-values for GDP and PI growth rate

Growth rate	$R_{89/87}$	$R_{91/89}$	$R_{93/91}$	$R_{95/93}$	$R_{98/95}$	$R_{00/98}$	$R_{02/00}$
$R_{\rm GDP}$	0.872	0.919	0.998	0.696	0.337	0.480	0.955
$R_{89/87}$		0.998	0.984	0.431	0.689	0.860	0.840
$R_{91/89}$			0.970	0.979	0.995	0.994	0.997
$R_{93/91}$				0.839	0.459	0.750	1.000
$R_{95/93}$					0.172	0.459	0.560
$R_{98/95}$						0.703	0.378
$R_{00/98}$							0.658

[7] See Lee *et al.* (1998), for similar findings about the GDP and company growth rates.

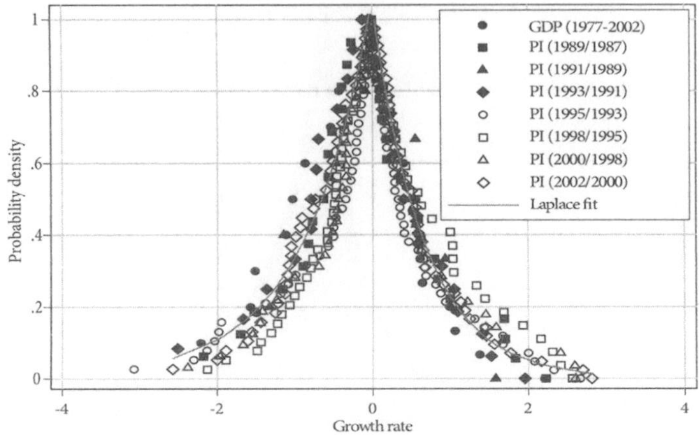

Fig. 3.15. Probability distribution of Italian GDP and PI growth rates

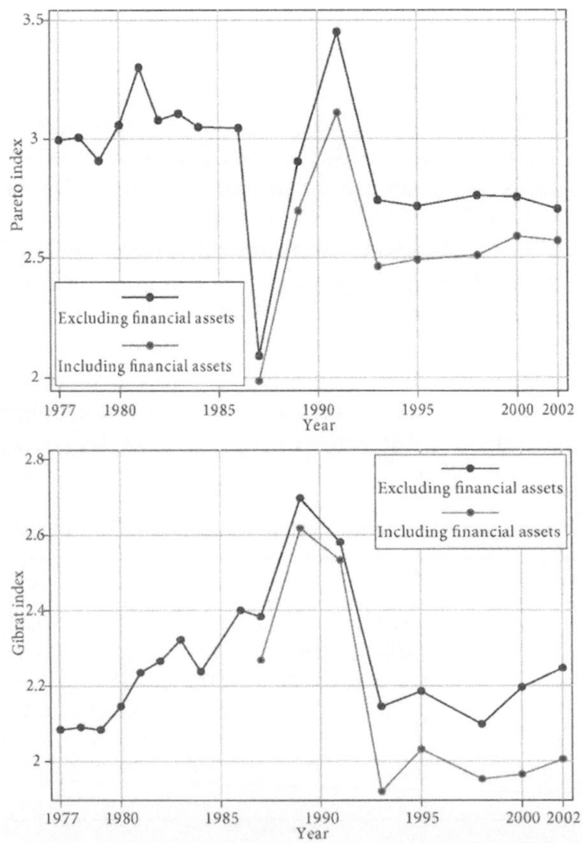

Fig. 3.16. The temporal variation of the Pareto (*left*) and Gibrat (*right*) indexes over the period 1977–2002

We now turn to the fluctuations of the indexes specifying the income distribution, i.e. the Pareto and Gibrat indexes, whose yearly estimates are reported in Fig. 3.16.

Since income deriving from financial assets has been regularly recorded only since 1987, we cannot take it into account for the period 1977–1987. The longest line in the graphs, therefore, depicts the time series obtained by excluding all the income deriving from financial assets, while the shortest one refers to the yearly estimates obtained from data including income from financial assets. The pattern of the two series is similar, with the more complete definition of income showing a greater inequality because of the strongly concentrated distribution of returns on capital.

Although the frequency of data (initially annual and then biennial from 1987) makes it difficult to establish a link with the business cycle, it seems possible to find a (negative) relationship between the Gibrat and Pareto indexes and the fluctuations of economic activity, at least until the late 1980s. For example, Italy experienced a period of economic growth until the late 1980s, but with alternating phases of the domestic business cycle: a slowdown of production up to the 1983 stagnation has been followed by a recovery in 1984, to be followed again by a slowdown in 1986. The values of Gibrat and Pareto indexes, inferred from the numerical fitting, tend to decrease in the periods of economic expansion (concentration goes up) and increase during the recessions (income is more evenly distributed).

The time pattern of inequality is shown in Fig. 3.17, which reports the change of the Gini coefficient over the considered period. The level of inequality decreased significantly during the 1980s and raised in the early 1990s, to be substantially stable in the following years. In particular, a sharp rise of the Gini coefficient (i.e., of inequality) is encountered in 1987 and 1993, cor-

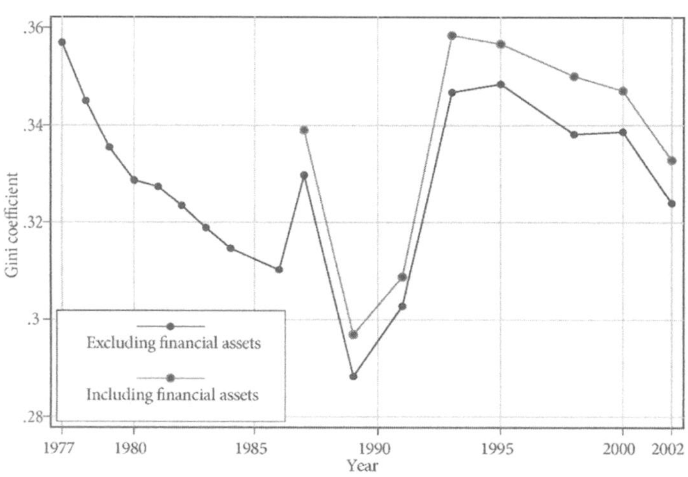

Fig. 3.17. The Gini coefficient for the Italian personal income, 1977–2002

responding to a sharp decline of the Pareto index in the former case and of both Pareto and Gibrat indexes in the latter case.

3.3 Power Law Scaling in the World Income Distribution

In the literature sprung up in recent years on the dynamics of the world distribution of per capita GDPs across countries, two empirical results have surfaced.[8] First, while convergence in terms of per capita GDP has been achieved among a restricted set of industrialized countries, i.e. the so-called *convergence club* (Baumol, 1986), divergence has been the rule for the GDP distribution taken as a whole (see e.g. Pritchett, 1997). Second, the density function of the cross-country GDPs distribution has moved from a unimodal shape in the 1960s to a "twin-peaks" shape in the 1990s (see e.g. Quah, 1993, 1996).

In this section we aim to add a new perspective to this literature very much in the spirit of the empirical methodology put forth so far, by discussing a third stylized fact regarding the world GDPs distribution which to our knowledge has been largely neglected so far.[9] We show that the GDP per capita of countries falling in the range between the 30th and the 85th percentiles of the distribution follows a power law, and that this result is extremely robust as we move from 1960 to 1997. Furthermore, over the same period the exponent of such a power law distribution displays a downward trend.

We study the world distribution of per capita GDPs as taken from the Penn World Table (PWT) Mark 6.1 (Summers, Heston and Ater, 2002), from 1960 to 1997. For the sake of brevity, in what follows we will refer to this object as the world income distribution. Though the PWT dataset contains estimates for some countries extending from 1950 to 2000, a restriction of the time horizon has been imposed in order to minimize the trade-off between the cross-section dimension and the time dimension of the panel.

Let the distribution of GDP per capita of M countries at year t be $\underline{\boldsymbol{x}}_t = (x_{1t}, \ldots, x_{Mt})$. Suppose each observation x_{it} is a particular realization of a random variable x with cumulative distribution function $F_t(x)$. Furthermore, let the observations in $\underline{\boldsymbol{x}}_t$ to be ordered from the largest to the smaller, so that the index i corresponds to the rank of x_{it}. When we make use of this simple algorithm to graphically represent the income distribution, which operationally corresponds to a scatter plot of the log of rank against the log of GDP per capita, we obtain a Zipf plot. As a matter of example, in Fig. 3.18 we show the Zipf plot of the world income distribution for $t = 1980$. Qualitatively similar findings hold for all the other years in our sample.

[8] Interesting reviews are e.g. Parente and Prescott (1993) and Jones (1997).

[9] For an example of work very close in spirit to ours, see Sinclair (2001).

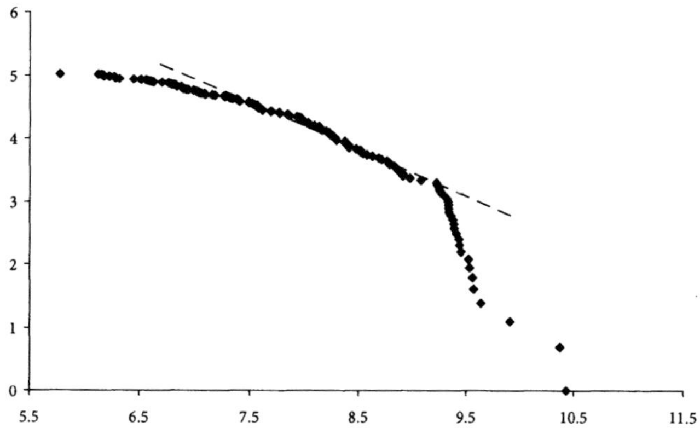

Fig. 3.18. Zipf plot of the world income distribution (GDP per capita) in 1980

In the figure we superimpose a dashed line, which helps us in visually isolating four different regions of the distribution: i) starting from the early 1970s, in several years there is a small group of extremely rich countries – typically, scarcely populated oil-producing ones – which can be considered outliers; ii) the remainder of the left tail consists typically of high income OECD countries, plus other more densely populated oil-producing nations; iii) the central part of the distribution, contains roughly 55% of the countries. For these countries, the log of per capita income is arranged along an interpolating line; iv) the right tail, which can be identified, for any practical purpose, with Africa.

The most intriguing feature emerging from this analysis is undoubtedly the regularity characterizing region iii), that is the fact that the data on GDP per capita for middle income economies fit a downward sloping straight line remarkably well. This fact holds invariably for the range of per capita GDP ranging from the 30th to the 85th percentiles[10] of the world income distribution in each single year of our time horizon, though the slope of the fitting line tends to change significantly over time, as one can easily recognize from Fig. 3.19.

We run an OLS regression for each year of the time span 1960–1997, for the data in the range between the 30th and the 85th percentiles of the world income distribution. The results are summarized in Fig. 3.20, where we plot the estimated value of the scaling exponent γ (continuous line)[11], and a measure of the goodness of fit expressed in terms of R^2 (dashed line).

The hypothesis that the central part of the world income distribution follows a power law seems to be corroborated by the extremely good fit of linear regressions, as one can appreciate by noting that the value of R^2

[10] This range has been obtained on a pure data-dependent basis.

[11] The coefficient γ was always statistically significant at the 1% level.

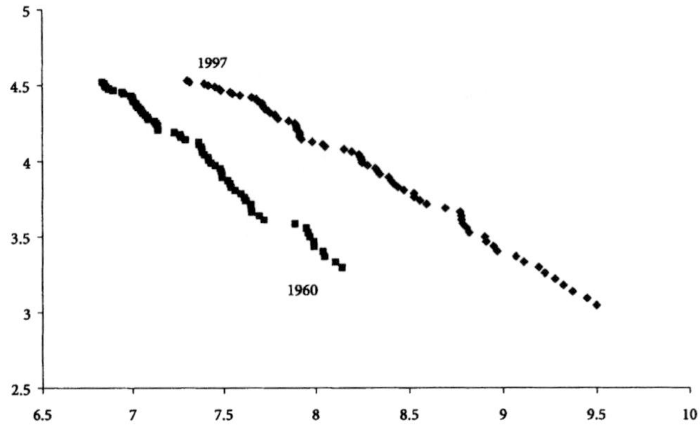

Fig. 3.19. Zipf plot of the 30th–85th percentiles of the world income distribution in 1960 and 1997

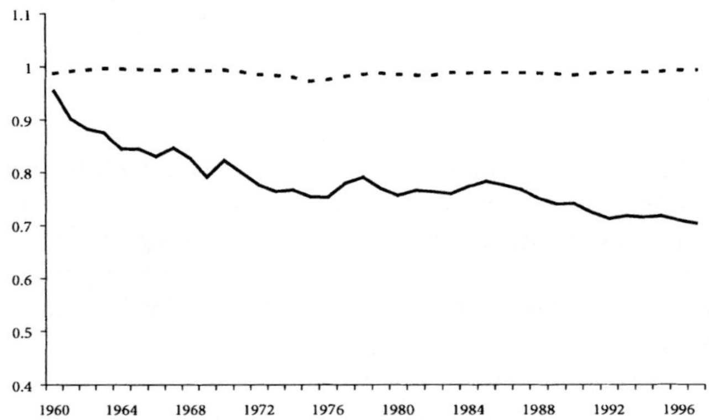

Fig. 3.20. Time path of the power exponent γ (*continuous line*), and goodness of fit of OLS estimates in terms of R^2 (*dashed line*)

is never below 0.978. Furthermore, note that γ shows a clear tendency to decrease over time. Both features have interesting implications for theory.

A possible explanation of this finding can be advanced along the following lines. Let us assume continuity both of GDP per capita levels and of time. Let $p(x, t; x_0)$ be the probability density function for x_t, where x_0 represents the initial condition. The evolution over time of $p(x, t; x_0)$ is given by the following Fokker–Plank diffusion equation:[12]

$$\frac{\partial p(x,t;x_0)}{\partial t} = -\frac{\partial \left[x\mu(x)\,p(x,t;x_0)\right]}{\partial x} + \frac{1}{2}\frac{\partial\left[x^2\sigma^2(x)\,p(x,t;x_0)\right]}{\partial x^2}, \qquad (3.7)$$

where $\mu(x)$ and $\sigma(x)$ are the drift and the diffusion coefficients, respectively. Cordoba demonstrates (Theorem 2, p. 14) that for the probability distribu-

[12] See Aoki 1996 for an introduction to the Fokker–Plank equation.

tion function of x to be Pareto with exponent γ, necessary conditions are that: i) the conditional mean, or drift, is constant, $\mu(x) = \phi$; ii) the diffusion coefficient takes the form $\sigma(x) = Ax^{\gamma-1}$, where A is a positive constant.

With reference to the issues at stake, these two conditions imply that countries belonging to the range of the world income distribution which scales down as a Pareto distribution are characterized by a common average growth rate ϕ, and that the variance of growth decreases with size as soon as $|\gamma| < 1$. The first condition, in particular, states a precise relationship between scale and growth, in that growth rates have to be scale-invariant.

This result is in line with the prediction of a recent stream of endogenous growth literature focusing on the driving role of R&D expenditure, according to which scale effects show up on GDP per capita levels, but not on growth rates[13]. Furthermore, the conjecture of a common average growth rate is consistent with panel data estimations provided by Evans (1998), who shows that the null hypothesis of different trend growth rates among a sample of countries with well-educated populations is rejected at standard statistical levels.

While steady state growth without scale effects seems to characterize countries with GDP per capita in the middle of the distribution, however, from our analysis it turns out that the mechanics of growth is likely to differ widely for very rich and very poor countries. In particular, the finding that growth processes for countries in the first 15% of the world income distribution seem to differ from those of the other high and middle income countries is somehow puzzling, and it deserves further research.

If the assumptions at the core of model (3.7) hold true, our estimates of γ imply that the variance of growth rates scales down on average as $\sigma^2(x) \sim x^{-0.22}$, meaning that the standard deviation follows a Pareto distribution with exponent $\beta = -0.11$. This guess is strikingly close to direct estimates of $\sigma(x)$ reported in Canning *et al.* (1998) and Lee *et al.* (1998), where $\beta = -0.15 \pm 0.03$. Notice that if an economy J consists of $j > 1$ identical and independently distributed units of size x_0[14], $x_j = jx_0$, the volatility of its growth tends to decrease with the square root of its size, so that for the whole vector $\underline{\boldsymbol{x}}$ fluctuations as a function of size should scale down with an exponent $\beta = -0.5$ (Buldyrev *et al.*, 1997). Therefore, an average β smaller (in absolute value) than 0.5 can be interpreted as suggesting the existence of long-range correlation between an economy's components, like in models of the business cycle based on direct interactions[15].

Furthermore, the negatively sloped trend of the estimated parameter γ signals that the volatility of fluctuations in countries in the lowest part of

---

13 See e.g. Dinopoulos and Thompson (1998) and Segerstrom (1998). Jones (1997) surveys the topic.

14 Think, e.g., to the multi-sector Real Business Cycle model of Long and Plosser (1983).

15 As a matter of example, the models by Durlauf (1996) and Aoki (1998).

the 30th–85th range of the distribution has been increasing in relative terms all over the span 1960–1997, so that β has actually increased over the same period. Of course, our analysis is unsuited to ascertain whether this fact is due to an increase in the amplitude of output fluctuations in low-income countries or to a decrease of volatility in countries with higher incomes. Independent evidence (Agenor *et al.*, 2000; IMF, 2001), however, seems to suggest that the first conjecture is likely to be the right one, probably reflecting a strengthening of the inverse relationship between income levels and vulnerability to financial and debt crisis.

3.4 Distributional Features of Aggregate Fluctuations

The last piece of evidence presented in this chapter is related to some distributional features of macroeconomic fluctuations, under the implicit assumption that the alternation of expansionary and contractionary phases of aggregate activity simply reflects the complex dynamics of firms' demography.

Our starting point consists in noticing the great effort that the profession has put forth so far to investigate whether the business cycle exhibits duration dependence. In fact, the received wisdom in mainstream macroeconomic theory is that business fluctuations are driven by recurring identically independently distributed (iid) random shocks, so that cycles' duration should be independent of length. The empirical work on duration dependence has been conducted by means of both nonparametric (McCulloch, 1975; Diebold and Rudebusch, 1990) and parametric techniques (Sichel, 1991; Diebold *et al.*, 1993; Zuehlke, 2003), the latter being generally favoured because of modeling convenience and higher statistical power.

While the evidence as a whole is far from conclusive, a relative consensus has been recently established in favour of positive duration dependence, at least for pre-war expansions and post-war contractions. A tentative explanation for this result has been advanced by Sichel (1991). Suppose that policymakers are interested in lengthening expansions. Hence positive duration dependence of contractions and null duration dependence of expansions might emerge simply because policymakers are urged to act countercyclically as downturns lengthens, while macroeconomic policy mistakes could be as likely to end short as well as long expansions.

This type of reasoning resembles the so-called stabilization debate, that is whether macroeconomic policy effectiveness in decreasing the volatility of business cycle fluctuations around trend has improved after the second World War (WWII). Romer (1999) argues in the affirmative while Stock and Watson (2005) have a partially dissenting view. Regardless of the position one is prone to take in this debate, it should be recognized that policymakers are plausibly not interested in the length of a business cycle phase (to end it if a recession, and to lengthen it if an expansion) per se. A recession could be disturbing not only if it is very long, but also if it is short *and* particularly severe. By the same token, an expansion could force antinflationary (i.e., countercyclical) policies if it is gaining excessive momentum, regardless of its duration. In other terms, macroeconomic policy could well be targeted at controlling the magnitude of business cycle phases, rather than their duration alone.

In line with this assumption, in this section we aim to extend the empirical literature on dependence in business cycles by posing a related but different question: are expansions or contractions in economic activity more likely to end as they become bigger?

The concept of business cycle fluctuations we adopt here is that of growth cycles, that is expansions and contractions expressed in terms of deviations from an estimated GDP trend or potential (Zarnowitz, 1992). A useful method to summarize information on either time (i.e., duration) and size (i.e. output gap) of any single episode consists in calculating its *steepness*,[16] expressed as the ratio of the amplitude y (i.e. cumulative percentage points of peak-to-trough and trough-to-peak output gap for recessions and expansions, respectively) to the amplitude t (in time periods), $x = \frac{y}{t} > 0$.[17] In what follows, we will use this measure to approximate the magnitude of a business cycle phase.

The following step consists in deriving the conditional probability that phase i will end at magnitude x_i, given that a magnitude x_i has been obtained. Our investigation is based on a Weibull parametric hazard model (Lancaster, 1979):

$$h_{\mathrm{W}}(x) = \alpha\beta x^{\beta-1} \tag{3.8}$$

and on its associated survivor function $S^{\mathrm{W}}(x) = \exp(-\alpha x^{\beta})$, with $\alpha = \eta^{-\beta}$, $\eta > 0$ being the *scale* parameter, while the *shape* parameter $\beta > 0$ measures the elasticity of magnitude dependence. If β is higher (lower) than one, then the conditional probability that a phase ends increases (decreases) linearly as its magnitude increases. Finally, when β is equal to 1, the hazard function corresponds to that of an exponential, or memory-less, distribution.

It is well known that in model (3.8) unobserved heterogeneity across observations biases the estimate of the elasticity parameter β downward (Lancaster, 1979). In particular, if estimates point towards negative magnitude dependence it could be practically impossible to identify whether such a result owes to true negative dependence or to heterogeneity bias. McDonald and Butler (1987) explain how to use mixture distributions to handle heterogeneity, showing that if the location parameter is inverse generalized gamma distributed, the distribution for observed data will be Burr type IIX.

---

16 The concept of steepness we use has a geometrical meaning and it is therefore different from the one in Sichel (1993), where this same term has been used to define a property of asymmetric business fluctuations.

17 This measure corresponds to the slope of the hypotenuse of the *triangle approximation to the cumulative movement* during a business cycle phase as discussed in Harding and Pagan (2002).

Our empirical exercise is based on a sample of pooled international data, so that heterogeneity is likely to seriously affect our estimates. Hence, in addition to the standard 2-parameter Weibull model (W) we check the robustness of our results by recurring to the hazard function for the generalized 3-parameter Weibull model proposed by Mudholkar *et al.* (1996) (MSK), which contains the Burr type XII distribution as a special case:

$$h_{\mathrm{MSK}}(x_i) = \alpha\beta x^{\beta-1}\left[S_{\mathrm{MSK}}(x_i)\right]^{-\gamma} \tag{3.9}$$

where $S_{\mathrm{MSK}} = [1-\gamma\alpha x^{\beta}]^{\gamma^{-1}}$is the corresponding survivor function, the *location* parameter γ is real, and the sample space for x is $(0,\infty)$ for $\gamma < 0$ and $(0,(\alpha\gamma)^{-(\beta-1)^{-1}})$ for $\gamma > 0$. Besides correcting for unobserved heterogeneity, the additional parameter γ allows the hazard function to be nonlinearly monotonic increasing ($\beta > 1$, $\gamma > 0$), nonlinearly monotonic decreasing ($\beta < 1$, $\gamma < 0$), bathtub shaped ($\beta < 1$, $\gamma > 0$), unimodal ($\beta > 1$, $\gamma < 0$) or constant ($\beta = 1$, $\gamma = 0$). Finally, when $\beta > 0$ and $\gamma \leq 0$ the generalized Weibull corresponds to the Burr type XII distribution.

For both models parameters' estimation has been conducted by means of Maximum Likelihood. The log-likelihood function for a series of expansions (contractions) with observed magnitude x_i is:

$$\ln L_j(\bullet) = \sum_{i=1}^{N}\left\{\Lambda_i \ln\left[h_j(x_i)\right] + \ln\left[S_j(x_i)\right]\right\} \tag{3.10}$$

with $j = \mathrm{W}, \mathrm{MSK}$, and where Λ_i is a binary variable that controls for the censoring of incomplete phases. Given that we are operating with nested models, a significantly positive (negative) estimate of γ is evidence, besides of positive or bathtub shaped (negative or unimodal) magnitude dependence, of a statistical superiority of the MSK model relative to the W model (Zuehlke, 2003). Furthermore, the sign of the estimated γ allows us to control for heterogeneity in the data: the magnitude elasticity obtained with the W model is likely to be biased downward as soon as γ in the MSK model is significantly negative.

The data we use are annual GDP indexes for 16 countries[18] spanning from 1881 through 2000 (IMF, 2002). The time series do not contain data for the periods corresponding to the two WWs, i.e. 1914–1919 and 1939–1948. For each country, the GDP potential has been calculated by means of the Hodrick–Prescott filter. Finally, in order to obtain enough observations to attain reliable estimates, we built samples for expansions (276 observations) and contractions (284 observations) by pooling data for individual countries.[19]

[18] The 16 countries are Australia, Canada, Denmark, Finland, France, Germany, Italy, Japan, Netherlands, Norway, Portugal, Spain, Sweden, Switzerland, United Kingdom and United States.

[19] This allows us to employ a number of observations one order of magnitude higher that the ones usually employed in the duration dependence literature.

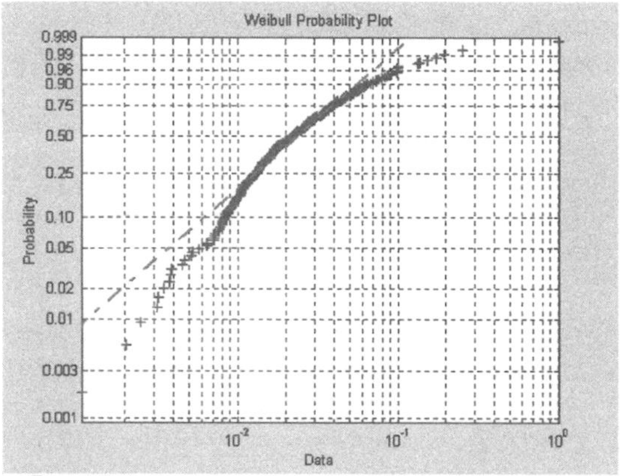

a Expansions

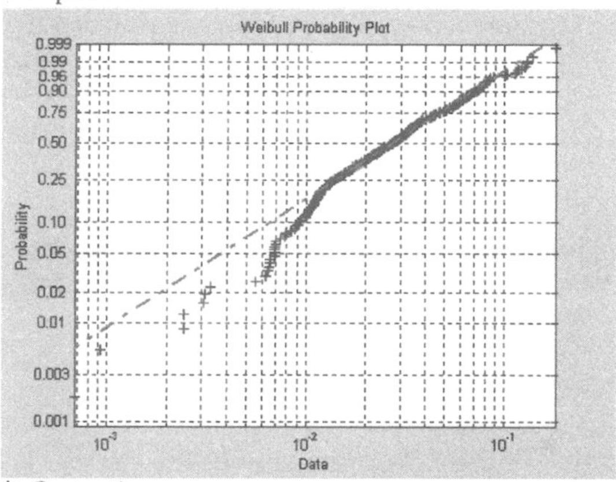

b Contractions

Fig. 3.21a,b. Weibull probability plots for expansions and contractions, full sample

Given that both parameterizations (3.8) and (3.9) yield hazard functions belonging to the Weibull family, as a preliminary check of model adequacy we study the magnitude empirical distribution of pooled expansions and contractions by means of Weibull probability plots. An advantage of such a plotting technique is that it allows to gain insights on the appropriateness of a distribution model by visual inspection: if the data come from a Weibull distribution the plot will be linear, while other distribution types would introduce curvature in the plot. Figure 3.21 shows that our modelling strategy finds a convincing support for contractions (Panel b), with observations distributed along the reference line but for few data points in the left tail, and a reasonable support for expansions (Panel a), with the Weibull model yielding a good fit to the empirical distribution for the central portion, but a relatively poor fit for both tails.

Table 3.4. Tests of magnitude dependence in pre-WWII, post-WWII and total sample expansions and contractions for a pool of 16 countries (1881–2000). Standard errors in *round brackets*, *p*-values in *square brackets*

	Expansions			Contractions		
	Total sample	Pre-WWII sub-sample	Post-WWII sub-sample	Total sample	Pre-WWII sub-sample	Post-WWII sub-sample
1) W model						
η	0.0387†	0.0517†	0.0259†	0.0401†	0.0533†	0.0265†
	(0.0024)	(0.0028)	(0.0027)	(0.0018)	(0.0024)	(0.0017)
β	1.0143	1.4440**	0.8643**	1.3832**	1.6792**	1.5173**
	(0.0384)	(0.0753)	(0.0444)	(0.0604)	(0.0912)	(0.1007)
2) MSK model						
η	0.0358†	0.0496†	0.0242†	0.0389†	0.0505†	0.0257†
	(0.0023)	(0.0028)	(0.0026)	(0.0018)	(0.0024)	(0.0017)
β	0.9635	1.3853**	0.8402**	1.3314**	1.5808**	1.4678**
	(0.0375)	(0.0730)	(0.0437)	(0.0586)	(0.0869)	(0.0979)
γ	0.0018†	0.0016	0.0011†	0.0009	0.0022	0.0006
	[0.0000]	[0.4091]	[0.0004]	[0.2259]	[0.4144]	[0.6995]

** Significantly different from unity at the 5% level using a one-tailed test.
† Significantly different from zero at the 5% level using a one-tailed test.

In Table 3.4 we report the Maximum Likelihood parameter estimates, along with their asymptotic standard errors, obtained with the W model (3.8) and the MSK model (3.9) for the full sample, the pre-WWII sub-sample and the post-WWII sub-sample, respectively. Expansions and contractions are treated separately.

The evidence from the W model shows that, for the total sample, positive magnitude dependence exists for contractions, while for expansions we are not able to reject the null of magnitude independence at standard significance levels with a one-tailed test. This last result occurs because of a structural change over the period studied. Pre-WWII expansions exhibit magnitude dependence ($\beta = 1.444$), while for the post-WWII sample the dependence elasticity is lower than one ($\beta = 0.8643$), meaning that in this case the probability expansions end decreases with their magnitude. In turn, contractions exhibit a substantially similar degree of positive magnitude dependence either in the pre-WWII ($\beta = 1.6792$) and in the post-WWII samples ($\beta = 1.5173$).

Estimates for the MSK model seems to confirm the robustness of our findings. The additional location parameter γ turns out to be positive but very small in all cases. In this case, an assessment of the statistical superiority of the MSK model relative to the W model cannot be based on standard tests build on asymptotic standard deviations because of their low power. Instead, a better strategy consists in using a *minimum statistic* test under the null

$\gamma = 0$, which returns the probability to observe a minimum above the ML estimate of γ.

Contractions still show positive magnitude dependence both in the pre-WWI and in the post-WWI era. As regards expansions, positive magnitude dependence is detected in the pre-WWII period, while for the post-WWII period the parameters estimates suggest a bathtub shaped hazard function. In fact, over the range of variation of our data the degree of non-linearity introduced by the MSK model is negligible for any practical purpose, as one can realise by visually inspecting the MSK hazard plots for expansions and contractions shown in Fig. 3.22.

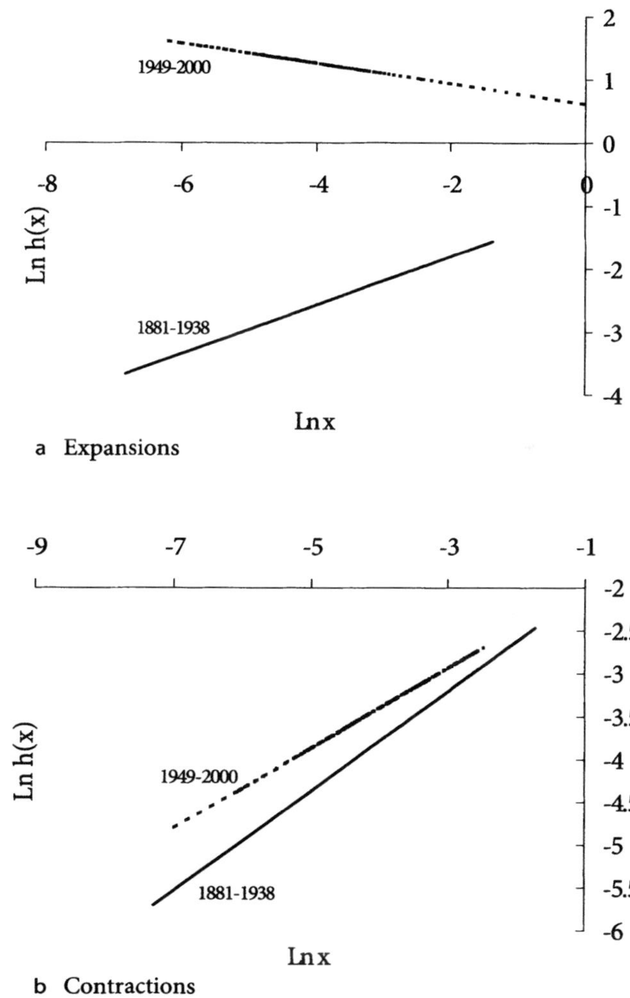

Fig. 3.22. Hazard plots for pre-WWII and post-WWII expansions and pre-WWII and post-WWII contractions

There are many possible explanations for the evidence at hand. Among them, the most appealing for us has to do with the so-called stabilization debate, that is whether the coming out of automatic stabilizers and the increased ability in conducting monetary policy after WWII has significantly contributed to decrease volatility in aggregate economic activity. Our starting point is that stabilization macroeconomic policy is generally aimed at affecting either the duration and the deepness of business cycle fluctuations: policymakers are better off if sustainable (i.e., without significant inflationary pressures) expansions lengthen a lot, and mild recessions are short. The measure we use for the magnitude of a business cycle phase, i.e. steepness, is designed precisely to capture both aspects. From this perspective, the structural shift we find for expansions before and after WWII could be interpreted as an indirect evidence that macroeconomic policy has became more effective from the 1950s on.

4 An Agent-based Model

4.1 Introduction

Reductionism, i.e. the methodology of classical mechanics which has been adopted by analogy in neoclassical economics, can be applied if the law of large numbers holds true, i.e.:

- the functional relationships among variables are linear; and,
- there is no direct interaction among agents.

Since non-linearities are pervasive, mainstream economics generally adopts the trick of linearizing functional relationships. Moreover agents are supposed to be all alike and not to interact. Therefore an economic system can be conceptualized as consisting of several identical and isolated components, each one being a representative agent (RA). The *optimal* aggregate solution can be obtained by means of a simple summation of the choices made by each optimizing agent.

Moreover, if the aggregate is the sum of its constitutive elements, its dynamics cannot but be identical to that of each single unit. The reductionist methodology implies that to understand the working of a system, one has to focus on the working of each single element. Assuming that elements are similar and do not interact – i.e. the economy is completely described by a representative agent – the dynamics of the aggregate replicate the dynamics of the sub-unit. The existence of an aggregate equilibrium, which is in turn "[...] *deeply rooted to the use of the mathematics of fixed point theory*" (Smale, 1976, p. 290), merely requires that every single element is itself in an equilibrium state.

The ubiquitous RA, however, is often at odds with the empirical evidence (Stoker, 1993)[1], is a major problem in the foundation of general equilibrium

[1] A modeling strategy based on the representative agent is not able, by construction, to reproduce the persistent heterogeneity of economic agents, captured by the skewed distribution of several industrial variables, such as firms' size, growth rates etc. Stoker (1993) reviews the empirical literature at disaggregated level which shows that heterogeneity matters since there are systematic individual differences in economic behavior. Moreover, as Axtell (1999, p. 41) claims: "... given the power law character of actual firms' size distribution, it would

theory (Kirman, 1992)[2,3] and is not perfectly coherent with many econometric investigations and tools (Forni and Lippi, 1997)[4]. All in all, we may say that macroeconomics (and macroeconometrics) still lacks sound microfoundations.

4.2 Heterogeneous Interacting Agents and Power Laws

The search for *natural laws* in economics does not necessarily require the adoption of the reductionist paradigm. Scaling phenomena and power law distributions are a case in point. If a scaling behavior exists, then the search for universality can be pushed very far. Physicists have shown that scaling laws are generated by a system with *strong* interactions and multiplicative shocks among heterogeneous agents (Marsili and Zhang, 1998; Amaral *et al.*, 1998) and therefore are incompatible with reductionism. As a consequence, the occurrence of scaling laws in economics is incompatible with mainstream economics. The macroscopic pattern (consisting of a multitude of heterogeneous interacting units) works as a unified whole independent of the dynamical process governing its individual components. The idea that systems which

seem that equilibrium theories of the firm [...] will never be able to grasp this essential empirical regularity."

2 According to Hildenbrand and Kirman (1988, p. 239): "... There are no assumptions on [...] isolated individuals which will give us the properties of aggregate behavior which we need to obtain uniqueness and stability. Thus we are reduced to making assumptions at the aggregate level, which cannot be justified, by the usual individualistic assumptions. This problem is usually avoided in the macroeconomic literature by assuming that the economy behaves like an individual. Such an assumption cannot be justified in the context of the standard economic model and the way to solve the problem may involve rethinking the very basis on which this model is founded." This long quotation summarizes the conclusion drawn by Arrow (1951), Sonnenschein (1972), and Mantel (1976) on the lack of theoretical foundations of the proposition according to which the properties of an aggregate function reflect those of the individual components.

3 In General Equilibrium theory one can put all the heterogeneity s/he likes, but no direct interaction among agents. Grossman and Stiglitz (1980) has shown that in this case one cannot have any sort of informational perfection. If information is not perfect markets cannot be efficient. Market failure leads to agents' interaction and to coordination failures, emerging properties of aggregate behavior, and to a pathological nature of business fluctuations.

4 If agents are heterogeneous, some standard procedures (e.g. cointegration, Granger-causality, impulse-response functions of structural VARs) may loose their significance. Moreover, neglecting heterogeneity in aggregate equations may generates spurious evidence of dynamic structure. The difficulty of testing aggregate models based on the RA hypothesis, i.e. to impose aggregate regularity at the individual level, has been long pointed out by Lewbel (1989) and Kirman (1992) with no impact on the mainstream (a notable exceptions is Carroll, 2000).

consist of a large number of interacting agents *generates* universal, or scaling, laws that do not depend on microscopic details is now popular in statistical physics and is gaining *momentum* in economics as well.

The quantum revolution of last century radically changed the perspective in contemporary physics, leading to a widespread rejection of reductionism. According to the holistic approach, the aggregate is different from the sum of its components because of the interaction of particles. The properties of the sub-units are not intrinsic but can be grasped only analyzing the behavior of the aggregate as a whole. The concept of equilibrium is therefore different from that of mainstream economics. The equilibrium of a system does not require any more that every element is in equilibrium, but rather that the aggregate is quasi-stable, i.e. in "[...] *a state of macroeconomic equilibrium maintained by a large number of transitions in opposite directions*" (Feller, 1957, p. 356).[5]

If the system is far from equilibrium, self-organizing phenomena and a state of self-organized criticality (SOC) may occur. According to the notion of SOC (Bak, 1997; Nørrelykke and Bak, 2002), scaling phenomena emerge because the sub-units of a system are heterogeneous and interact, and this leads to a *critical state* without any attractive point or state[6]. Since scaling phenomena characterize such critical states, the occurrence of a power law may be read as a symptom of self organizing processes at work. A notable example of this approach applied to macroeconomics is the inventory and production model developed by Bak *et al.* (1993).

If a distribution is described by a power law, firms are located along a curve whose coefficient is stable and the intercept changes very slowly over time.[7] This is due to the fact that the data generating process is random: in terms of the states of a dynamics process we may say that the transition from one state to another is affected by *chance* as well by agents' *systematic actions.*[8]

---

[5] Moreover agents' choice should not necessarily be an equilibrium one, derived from their optimizing behavior, because agents' interaction generates self-organizing solutions. It follows from this that one should not analyze the individual problem in isolation from the others (a game *against nature*) but rather the interconnections among HIAs.

[6] In the SOC literature the concept of equilibrium is borrowed from *statistical mechanics* and is very different from that of mainstream economics. In fact, equilibrium results from the balance of actions of a large number of many interacting particles.

[7] Stability of the slope through time is a quite standard result in the empirical literature on Pareto's law (see e.g., the work by C. Gini, J. Steindl and H. Simon). Quite nicely, Steindl (1965, p. 143) defines the Pareto coefficient "... a sediment of growth over a long time".

[8] The biased behavior of this *random process* helps to explain the systematic differences (*asymmetries*) between expansions and contractions found in the empirical

In the model of Sects. 4.5 and 4.6, output fluctuations are due to: 1) a random process on current revenues as a consequence of imperfect information on actual prices; 2) systematic interactions among agents. The distribution is quasi-stable over relatively long periods because it represents "[...] *slowly changing, age-dependent characteristics of a population which ages and renews itself only gradually*" (Steindl, 1965, p. 142). This means that, since firms are born small, their growth takes time and mortality decreases with age and size, the slow change of the distribution comes as a consequence. In a nutshell: the distribution is stable, or quasi-stable, because the dissipative force of the process (here, the Gibrat's law) produces a tendency to a growing dispersion, which is counteracted by a stabilizing force (i.e., the burden of debt commitments and the associated risk of bankruptcy).

Moreover, distributions are interconnected. The population is characterized by a joint distribution of several variables (in our model: equity, capital, debt, age, equity ratio), which is inconsistent with the RA framework. The change of firms' distribution (and the business cycle itself) has to be analyzed in terms of changes of the joint distribution of the population.[9]

Alternatively, and in some sense in a way more germane to the economics discourse, power laws can be generated by models based on scale free growth processes. The basic idea can be traced back to the well-known Simon's model (Simon, 1955), where the Gibrat's law of proportional effects is combined with an entry process to obtain a Lévy distribution for firms' size. Furthermore, recent work by physicists (e.g. Marsili and Zhang, 1998; Amaral *et al.*, 1998) has shown that, by extending the heterogeneity of the system's components implied in Simon's scheme to account for direct or indirect interactions among units, power laws emerge naturally and, most notably, without the disturbing asymptotic implications of the original Simon's model or of its modern successors, like the one by Gabaix (1999).[10]

It is worthwhile to stress that, regardless of the modeling strategy one chooses, the adoption of the scaling perspective in economics implies rejecting the very definition of a representative agent because the dynamics of the system originate from the interaction of heterogeneous agents. We believe that, in order to grasp the empirical evidence and provide a coherent framework, economists have to adopt a methodological approach based on heterogeneous interacting agents (HIA).

---

evidence. Gaffeo *et al.* (2003) have found systematic differences of the Pareto exponents during expansions and contractions.

9 A recession, e.g., is more likely when firms are relatively young, small and leveraged. The RA framework not only is inconsistent with the evidence (i) but it also misses and outguesses any dynamical properties of the actual systems (Forni and Lippi, 1997).

10 Discussions on this point can be found in Krugman (1996) and Blank and Solomon (2000).

4.3 Agent Based Modeling

A step in this direction is the agent-based modeling strategy, which is increasingly applied also in economics (Epstein and Axtell, 1996; Tesfatsion, 2002). At the simplest level, agent-based models are computer programs that simulate the autonomous behavior of individual entities and the relationships between them. Such virtual environments are particularly powerful and flexible, as they can be employed for advancing theoretical conjectures as well as for testing alternative normative prescriptions in a controlled situation. In fact, we claim that the agent-based approach represents a fruitful methodology to do *realistic* macroeconomics, that is one based on bounded rational, heterogeneous interacting agents adapting to a complex world.

In a sense, agent-based computational techniques provide a route to develop microfoundations for macroeconomics completely at odds with the RA approach. The relevance and reliability of these new microfoundations are grounded in the empirical evidence they can account for. From this viewpoint, microfoundations can be defined as sound if they are based on a reasonable model of individual behavior and market and non-market interactions, who in the aggregate can produce regularities consistent with the empirical evidence, instead of being grounded on optimizing principles and equilibrium solutions.

In our approach the origin of business fluctuations – which is the most important single problem in macroeconomics – can be traced back to the ever changing configuration of the network of heterogeneous interacting firms.[11] A major role in shaping dynamics is played by financial variables. The sequential timing structure of our economy implies that future markets are absent, so that agents have to rely on means of payment – here, bank credit extended to firms – to bridge the gap between agents' decisions and their realization. Highly leveraged (i.e., financially fragile) firms, in turns, are exposed to a high risk of default, that is of going bankrupt. When bankruptcies occur, loans not refunded negatively affect banks' net worth, with banks responding to their worsen financial position by reducing credit supply. The reduction in credit supply impacts on the lending interest rate all other firms have to pay to serve their financial commitments.

In what follows, we build on the HIA framework developed in Gallegati *et al.* (2003) and Delli Gatti *et al.* (2005) to put at work all the notions we surveyed in this section, by modeling an economy characterized by aggregate scaling behaviors due to multiplicative idiosyncratic shocks and interactions among firms.

[11] Schumpeter (1939) suggested that business cycle scholars should analyze "... how industries and individual firms rise and fall and how their rise and fall affect the aggregates and what we call loosely general business conditions". This approach is reminiscent of Marshall's parallel between the dynamics of the individual firm and the evolution of a tree in the forest.

In Chap. 2, we observed how data on industrial dynamics display several empirical regularities which emerge so neatly across countries and over time to be characterized as *empirical* laws. Here we focus on two of them, that is: (i) the distribution of firms' size is right skew and can be described by a Zipf or power law probability density function; (ii) the growth rates of firms' output and countries' GDP follow a Laplace distribution.

So far, the literature has generally dealt with (i) and (ii) as if they were independent stylized facts. In this chapter we aim at making three contributions.

- We explore the link between the two, showing that the *power law distribution of firms' size* may be at the root of the growth rate fat tail distribution observed in empirical data. In this part we analyze also the exact conditions under which an *exact* power law distribution of firms' size implies double exponential rate of growth.
- We discuss a model of *financial fragility*, empirically validated through conditioning (Brock, 1999), which generates fact (i) (Sect. 4.5).
- We show that the features of *business fluctuations* such as the shifts of the distribution of firms' size over the cycle, the properties of the distribution of individual and aggregate growth rates and many others, are a consequence of (i) (Sect. 4.6).

While the industrial organization literature has explored at length the regularities (i) and (ii) at least since the 1950s,[12] inadequate attention has been paid so far to establishing a link between them and business cycle theory. We argue that this is mainly because mainstream macroeconomics lacks adequate conceptual and analytical tools to accomplish such an endeavor.

4.4 Scale Free Property and Fat Tails Rate of Growth

In this section we analyze the relationship between size and the rate of growth of firms. First, we show that simple multiplicative processes with *Gaussian shocks* generating firms' size with Pareto shape may also generate rate of growths with fatter than Gaussian tails. Second, we investigate the exact conditions under which rate of growth with double exponential shape are derived.

The simplest modification of the Gibrat's multiplicative process (see Sect. 2.2), able to generate power laws distributions, is the Kesten's process (Kesten, 1973) in which there is a minimum firm's size, say S_m. The process generates a Pareto distribution with exponent near to unity.

In Fig. 4.1 is depicted the rate of growth (computed as log-differences of firm's size in the last two iterations) excess kurtosis (i.e., kurtosis minus 3)

[12] For a review of the debate on the shape of the firms' size distribution sprung up during the 1950s and '60s, see the monograph by Steindl (1965).

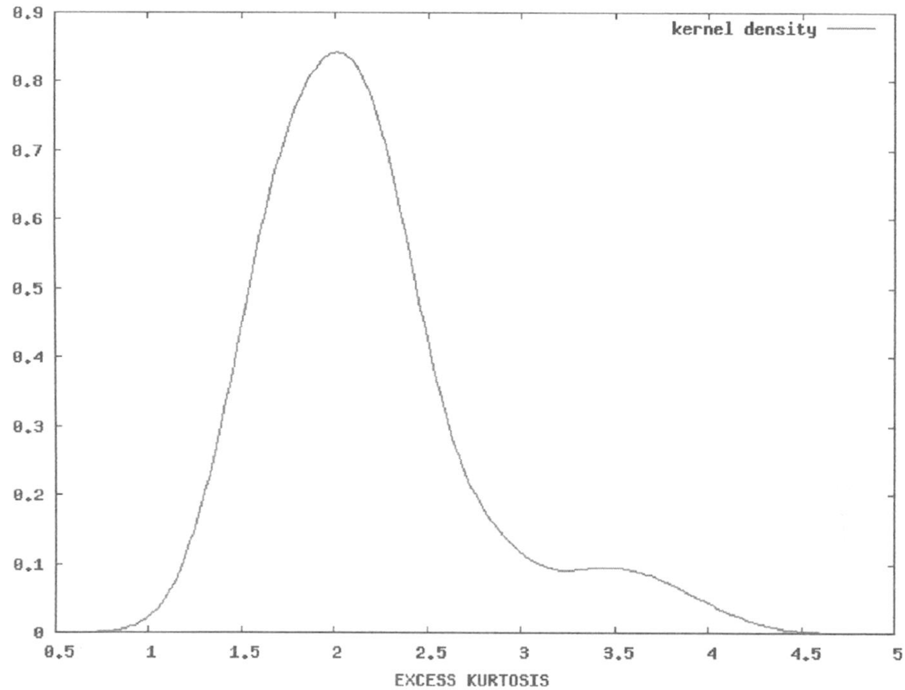

Fig. 4.1. Excess kurtosis distribution for the Kesten's process

distribution of 100 simulations with $S_m = 100$, initial size equal to 1000, and 10,000 iterations. The multiplicative shocks are extracted from a Gaussian distribution with standard deviation equal to 1/4.

As shown in the figure, the process – with a sufficiently high level of individual volatility – generates fatter than Gaussian distributions (whose excess kurtosis is zero). An example of such distributions is drawn in Fig. 4.2. However, the distribution is not well approximated by a double exponential in whole support and seems to be asymmetric.

This is more clearly shown in Fig. 4.3 where we compute the *qqplot* for the standardized rate of growth (i.e., the rate of growth are transformed in order to have the mean and variance of a standard Laplace centered in zero and scale parameter equal to 1). The Laplace model seems to be a useful approximation of the rate of growth generated by the Kesten's process only in the center-left part of the support. Furthermore such process generates a positive probability mass at zero. In other terms, contrary to the standard Laplace distribution there is a positive probability (about 7% in the simulation) that firms do not change their size.

The exact conditions under which *exact* Pareto distribution for firms' size implies double exponentials rate of growth are discussed in Palestrini (2007). One can start from the definition of the growth rate as the log-difference of

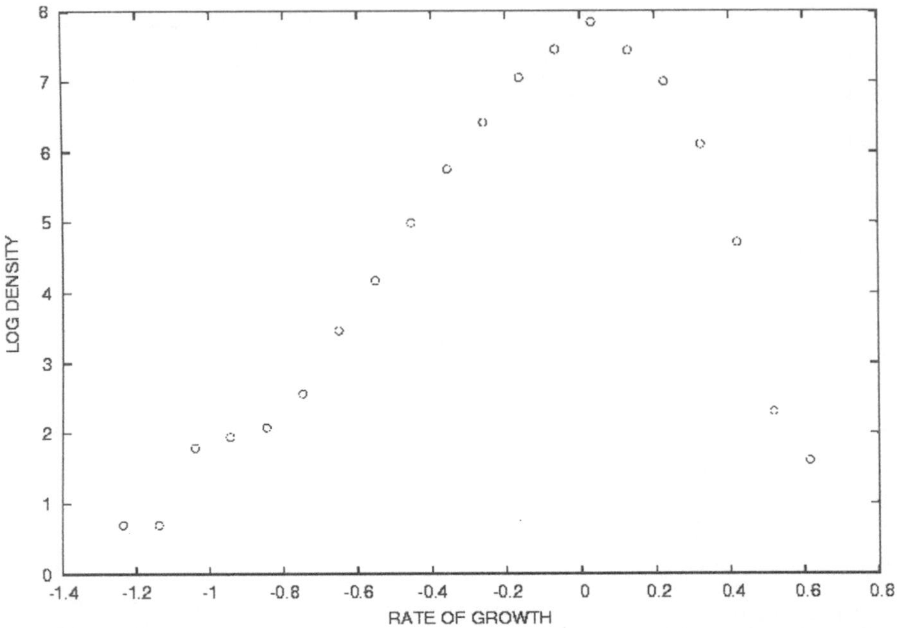

Fig. 4.2. Density estimation of a Kesten's process in the semi-log plane

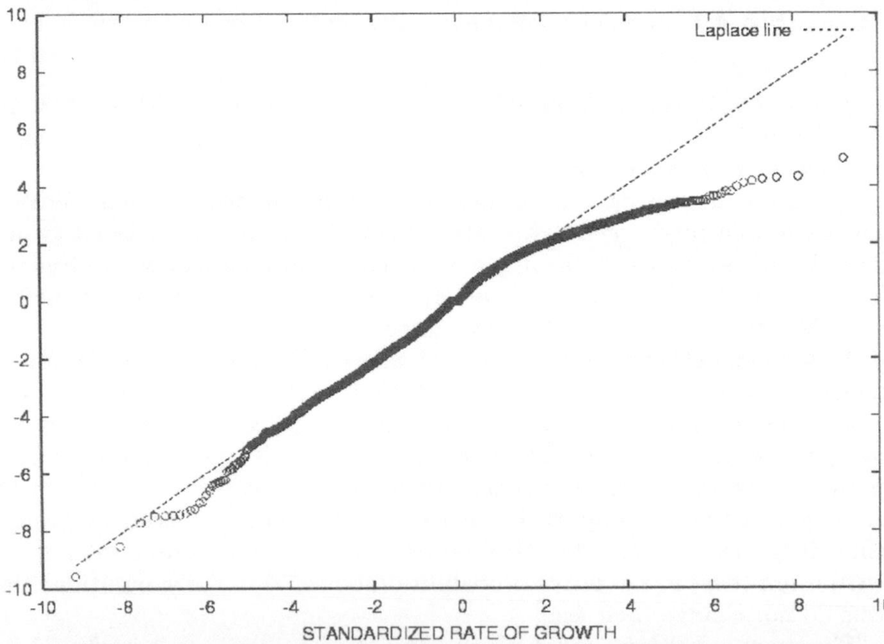

Fig. 4.3. *qqplot* for the standardized rate of growth generated by a Kesten's process

the state variable's levels, so that the proof consists in showing that: 1) the logarithm of a Pareto random variable follows an exponential distribution; and 2) the difference of two exponential random variables becomes a double exponential distribution.

Let us start by assuming that the distribution of the firms' size is Pareto. The most immediate way to think at the statistical properties of firms' growth at any t consists simply in exploiting the statistical features of firms' sizes at consecutive periods of time. If we define, as in the simulation, the growth rate g_t of size S_t as the log-difference between sizes at time t and $t-1$:

$$g_t = \log(S_t) - \log(S_{t-1}) , \tag{4.1}$$

it is immediate to note that, without any further information on the firms' size joint density function, g_t represents the difference of two dependent exponential distributions. The proof of this proposition is based on the monotonic property of the logarithmic function and on the rule of transformation of random variables.

Let the random variable S to follow a Pareto distribution with parameter α. Thus, the probability distribution of $\log(S)$ is:

$$\Pr(\log(S) \geq k) = \Pr(S \geq \exp(k)) \propto (\exp(k))^{-\alpha} = \exp(\text{-}\alpha k) , \tag{4.2}$$

that is, an exponential distribution with parameter α. In other terms, $\log(S)$ follows an exponential distribution with probability density function equal to:

$$E\left(\log(S); \alpha^{-1}\right) = \frac{1}{\alpha} \exp\left(-\frac{\log(S)}{\alpha}\right) . \tag{4.3}$$

In the case of *independent exponential variables*, it is simple to prove that an *exact* Laplace distribution regarding growth rates emerges by making use of the convolution theorem and its relation with the characteristic function. In fact, the characteristic function of two independent exponential distributions z_j, $j = 1, 2$, with parameter α^{-1} is:

$$C_{z_j}(\gamma) = (1 - i\alpha\gamma)^{-1} , \tag{4.4}$$

while their difference $y = z_1 - z_2$ has a characteristic function that is the product of the two, that is:

$$C_y(\gamma) = C_{z_1}(\gamma) C_{z_2}(-\gamma) = (1 - i\alpha\gamma)^{-1} (1 + i\alpha\gamma)^{-1} = \left(1 + \alpha^2\gamma^2\right)^{-1} , \tag{4.5}$$

that is the characteristic function of a Laplace distribution.

To prove the existence of a relationship between the Pareto distribution for firms' size and the double exponential like distribution for their growth rates in the case of time dependent exponential distributions, first note that for any exponential distribution s (that, for what said above, may be thought as the log of S) the two following properties hold true:

$$\Pr\left(s_t > \bar{s}_1 + \bar{s}_2 \,|s_t > \bar{s}_1\right) = \Pr\left(s_t > \bar{s}_2\right) , \tag{4.6}$$

$$\Pr\left(s_{t+1} > \hat{s}_1 + \hat{s}_2 \,|s_{t+1} > \hat{s}_1\right) = \Pr\left(s_{t+1} > \hat{s}_2\right) . \tag{4.7}$$

Marshall and Olkin (1967) proved that whenever the two properties above also hold true (in a sense) for the joint probability distribution of s_t and s_{t+1}, that is:

$$\Pr\left(s_t > \bar{s}_1 + k, s_{t+1} > \hat{s}_1 + k \,|s_t > k, s_{t+1} > k\right) = \Pr\left(s_t > \bar{s}_t, s_{t+1} > \hat{s}_t\right) , \tag{4.8}$$

then the only bivariate exponential distribution function consistent with (4.8) which has exponential marginals is given by:

$$\Pr\left(\log\left(S_t\right) > s_1, \log\left(S_{t+1}\right) > s_2\right) = \exp\left(-\alpha_1 s_1 - \alpha_2 s_2 - \lambda \max\left(s_1, s_2\right)\right) , \tag{4.9}$$

where λ is a measure of dependence. In other terms, $\lambda = 0$ means independence whereas when $\lambda > 0$ there is a positive dependence between observations in t and $t+1$. A well known result of this bivariate distribution is that is not absolutely continuous (i.e., has a positive probability mass for the case $s_1 = s_2$).

If the growth rate is positive ($g = s_2 - s_1 > 0$) the joint density f of (s_1, s_2) is proportional to the following exponential:

$$f\left(s_2, s_1\right) \propto \exp\left(-\alpha_1 s_1 - \alpha_2 s_2 - \lambda s_2\right) . \tag{4.10}$$

Integrating along the line $s_2 = s_1 + g$ in the plane (s_1, s_2) allows us to obtain the following relation for the probability density function of g, $\phi(g)$:

$$\begin{aligned} \phi\left(g\right) &\propto \int_{s_1=0}^{\infty} \exp\left(-\alpha_1 s_1 - \left(\lambda + \alpha_2\right) s_1 - \left(\lambda + \alpha_2\right) g\right) \mathrm{d}s_1 = \\ &= \exp\left(-\left(\lambda + \alpha_2\right) g\right) \int_{s_1=0}^{\infty} \exp\left(-\alpha_1 s_1 - \left(\lambda + \alpha_2\right) s_1\right) \mathrm{d}s_1 , \end{aligned} \tag{4.11}$$

that implies:

$$\phi\left(g\right) \propto \exp\left(-\left(\lambda + \alpha_2\right) g\right) . \tag{4.12}$$

By symmetry, the probability density function of $|g|$ for the case $g < 0$ satisfies the following condition:

$$\phi\left(|g|\right) \propto \exp\left(-\left(\lambda + \alpha_1\right) |g|\right) . \tag{4.13}$$

Relations (4.12) and (4.13) show that the rate of change of firms size follows a double exponential like distribution. Technically, such distribution is not Laplace since, as discussed in Bottazzi (2007), has a positive probability that firms do not change their size. This properties comes from the Marshall–Olkin bivariate distribution that is not absolutely continuous. Bottazzi (2007) also note that the double exponential like distribution derived in Palestrini (2007) does not fit empirical data. An explanation (Palestrini, 2007) is that such distribution is very sensitive to the conditions under which it is derived. The Marshall–Olkin condition and, as said before, power law distributions of firms' size (and also the Laplace model for firms' growth rate) are not good approximations for the whole support of the respective empirical distributions.

The main message, for an economist, looming large from the statistical results discussed so far is that the scaling approach to business fluctuations may derive in the first place from the levels of state variables being distributed as a power law, at least qualitatively in the sense that may generate fat tail distributions. Thus, the basic question to be answered is whether scale invariance for levels of state variables, in a certain range of the support, is a general feature of economic systems or not. From this viewpoint, it emerges that power law probability functions arise endogenously in economics basically for two reasons: 1) the lack of a characteristic scale in empirical and theoretical economics, implying that the occurrence of either rare or frequent events (i.e., sizes) is governed by the same law (Zajdenweber, 1997); 2) a power law behavior in the tail(s) of a distribution is a feature of a family of distributions known as *Lévy-stable distributions.* Due to a generalization of the central limit theorem (Gnedenko and Kolmogorov, 1954), the sum of a large number of identical and independent random variables has a probability density function characterized by a four-parameter characteristic function. Among the many different available parameterizations, we choose the $S_0(\alpha, \beta, \gamma, \delta)$ parameterization proposed by Nolan (2002),[13] according to which the characteristic function of X is given by:

$$E\exp(\mathrm{i}tX) \tag{4.14}$$
$$= \begin{cases} \exp\left\{-\gamma^{\alpha}|t|^{\alpha}\left[1+\mathrm{i}\beta\left(\tan\frac{\pi\alpha}{2}\right)(\operatorname{sign}t)\left((\gamma|t|)^{1-\alpha}-1\right)\right]+\mathrm{i}\delta t\right\} & \text{if} \quad \alpha\neq 1 \\ \exp\left\{-\gamma^{\alpha}|t|^{\alpha}\left[1+\mathrm{i}\beta\frac{2}{\pi}(\operatorname{sign}t)(\ln|t|+\ln\gamma)\right]+\mathrm{i}\delta t\right\} & \text{if} \quad \alpha=1\,. \end{cases}$$

A major advantage of the functional form (4.14) is that the four parameters have an intuitive interpretations. The characteristic exponent or index of stability α, which has a range $0 < \alpha \leq 2$, measures the probability weight in the upper and lower tails of the distribution. In general, the p-th moment of

[13] The parameterization of the characteristic function S_0 is particularly convenient because the density and the distribution functions are jointly continuous in all four parameters.

a stable random variable is finite if and only if $p < \alpha$. Thus, for $\alpha < 2$, a Lévy-stable process possesses a mean equal to the location parameter δ (which in turn indicates the centre of the distribution) but it has infinite variance, while if $\alpha < 1$ even the mean of the distribution does not exist. β, defined on the support $-1 \leq \beta \geq 1$, measures the asymmetry of the distribution, with its sign indicating the direction of skewness. Finally, the scale parameter γ, which must be positive, expands or contracts the distribution around the scale parameter δ. The Lévy-stable distribution function nests several well-known distributions, like the Gaussian $N(\mu, \sigma^2)$ (when $\alpha = 2$, $\beta = 0$, $\gamma = \sigma^2/2$ and $\delta = \mu$), the Cauchy ($\alpha = 1$ and $\beta = 0$) and the Lévy–Smirnov ($\alpha = 0.5$ and $\beta = \pm 1$).

Lévy-stable distributions are particular important because they represent an attractor in the functional space of probability density functions, in that the generalized Central Limit Theorem (Gnedenko and Kolmogorov, 1954) states that the only possible limiting distribution for sums of independently and identically distributed random variables belongs to the Lévy-stable family. It follows that the conventional Central Limit Theorem is just a special case of the above – a special case which applies whenever one imposes the condition that each of the constituent random variables has a finite variance. In particular, Lévy-stable distributions are stable under convolution. Simply stated, if we sum N *iid* Lévy-distributed variables with characteristic exponent α:

$$T_N = \frac{\sum_{i=1}^{N} \tau_i}{N^{\frac{1}{\alpha}}} , \tag{4.15}$$

the renormalized sum T_N is also Lévy-stable with characteristic exponent α. Besides other interesting properties,[14] non-Gaussian Lévy-stable distributions (i.e., for $\alpha < 2$) are characterized by tails which are asymptotically Pareto distributed with exponent $1 + \alpha$.

When $\alpha = 2$, $\beta = 0$ and $\gamma = \sigma^2/2$, the distribution is Gaussian with mean μ and variance σ^2. The Gaussian family is the only member of the Lévy class for which the variance exists. The presence of second moments implies that, if disturbances hitting firms are only idiosyncratic ones, aggregate fluctuations disappear as the number of firms N grows large. In fact, without aggregate shocks the variance of the average output of N firms is less than the maximum variance of firms' output, say $\sigma^2_{\max}/N$, a quantity that, for N going to infinity, vanishes. On the contrary, stable distributions with $\alpha < 2$ do not need aggregate shocks to generate aggregate fluctuations.[15,16]

The difference between the two situations may be well described by the example depicted in Fig. 4.4, where we report the average output time path of

[14] A comprehensive reference is Samorodnitsky and Taqqu (1994).
[15] The first author conjecturing it explicitly has been Mandelbrot (1960).
[16] From this viewpoint, in real world normality (in terms of Gaussian distributed variables) might just be a special case.

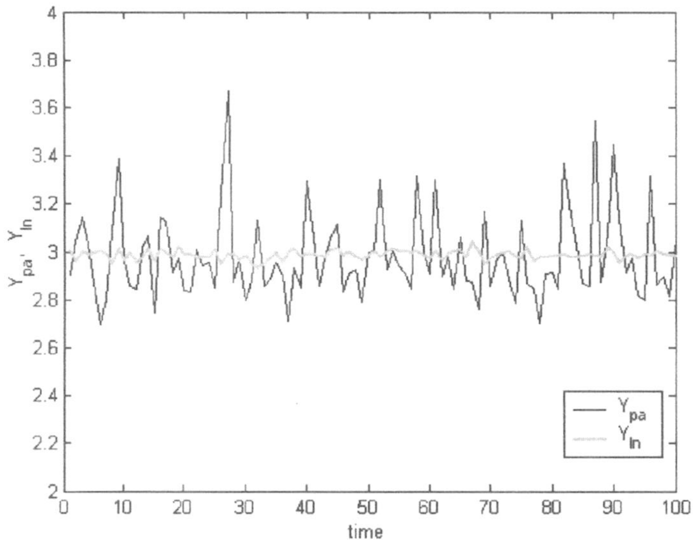

Fig. 4.4. Comparison between two simulated economies, inhabited by 10,000 firms each. In the first economy (*black line*) agents' size distribution is Pareto with location parameters $k = 1$ and stability parameter $\alpha = 1.5$. In the second economy (*grey line*) agents' size distribution is lognormal, with the same mean (i.e., 3) and same estimated variance at $t = 0$ (i.e., 10.4) of the other economy. The plot describes the two time series of agents' average output, Y_{pa} (the time evolution of the mean of the Pareto distributed firms) and Y_{ln} (the time evolution of the mean of the lognormal distributed firms)

two economies identical but for the shape of their firms' size distributions. In the first economy firms are Pareto distributed, whereas in the second one the distribution of firm's size is lognormal. Output is assumed to be proportional to size. Time series have been obtained by averaging from samples extracted from the two distributions at any time period t.

The time evolution of the average output shows almost no aggregate fluctuations for the lognormal economy, but large fluctuations in the Pareto economy even in the absence of aggregate shocks. In particular, the variance of the average aggregate output in the Pareto case is one order of magnitude greater than the variance of the lognormal case. Put differently, stable Pareto–Lévy distributions are good candidates to explain aggregate large fluctuations in time periods characterized by small aggregate shocks.

4.5 An Agent-based Model

Consider a sequential economy,[17] with time running discretely in periods $t = 1, 2, \ldots$, populated by many firms and banks. Two markets are opened in each period: the market for an homogenous produced good, and the market for credit. As in the *levered aggregate supply* class of models first developed by Greenwald and Stiglitz (1990, 1993), our model is fully supply-determined,[18] in the sense that firms can sell all the output they (optimally) decide to produce.

Due to informational imperfections on the equity market, firms can raise funds only on the credit market. The demand for credit is related to investment expenditure, which is therefore dependent on banks' interest rates. Total credit supply, in turn, is a multiple of the banks' equity base, which is negatively affected as insolvent borrowing firms go bankrupt. As we will discuss below, this mean-field interaction provides a mechanism to create long-range inter-temporal correlations capable to amplify and propagate idiosyncratic shocks.

4.5.1 Firms

At any time period t, the supply side of the economy consists of finitely many competitive firms indexed with $i = 1, \ldots, N_t$, each one located on an island. The total number of firms (hence, islands) N_t depends on t because of endogenous entry and exit processes to be described below. Let the i-th firm uses capital (K_{it}) as the only input to produce a homogeneous output (Y_{it}) by means of a linear production technology, $Y_{it} = \phi K_{it}$. Capital productivity (ϕ) is constant and uniform across firms, and the capital stock never depreciates.

The demand for goods in each island is affected by an *iid* idiosyncratic real shock. Since arbitrage opportunities across islands are imperfect, the individual selling price in the i-th island is the random outcome of a market

---

[17] Recall that in a sequential economy (Hahn, 1982) spot markets open at given dates, while future markets do not operate.

[18] Two scenarios are consistent with this assumption. In the *equilibrium* scenario, aggregate demand accommodates supply, i.e. households and firms absorb all the output produced by the latter and the goods market is always in equilibrium. In this scenario, aggregate investment must be equal to the sum of retained profits and households' saving. As we will see, both investment and retained profit are determined in the model, so that we have to assume that households' saving adjusts in order to fill the gap between the two. In the *disequilibrium* scenario, aggregate demand does not (necessarily) accommodate supply, so that the goods market is generally not in equilibrium. In this case, the difference between aggregate investment on the one hand, and the sum of profit and households' saving on the other must be assumed to take the form of involuntary inventories decumulation.

process around the average market price of output P_t, according to the law $P_{it} = u_{it} P_t$, with expected value $E(u_{it}) = 1$ and finite variance.

By assumption, firms are fully rationed on the equity market, so that the only external source of finance at their disposal is credit. The balance sheet identity implies that firms can finance their capital stock by recurring either to net worth (A_{it}) or to bank loans (L_{it}), $K_{it} = A_{it} + L_{it}$. Under the assumption that firms and banks sign long-term contractual relationships, at each t debt commitments in real terms for the i-th firm are $r_{it} L_{it}$, where r_{it} is the real interest rate.[19] If, for the sake of simplicity, the latter is also the real return on net worth, each firm incurs financing costs equal to $r_{it}(L_{it} + A_{it}) = r_{it} K_{it}$. Total variable costs proportional to financing costs[20], $g r_{it} K_{it}$, with $g > 1$. Therefore, profit in real terms (π_{it}) is:

$$\pi_{it} = u_{it} Y_{it} - g r_{it} K_{it} = (u_{it}\phi - g r_{it}) K_{it} , \tag{4.16}$$

and expected profit is $E(\pi_{it}) = (\phi - g r_{it}) K_{it}$.

In this economy, firms may go bankrupt as soon as their net worth becomes negative, that is $A_{it} < 0$. The law of motion of A_{it} is:

$$A_{it} = A_{it-1} + \pi_{it} , \tag{4.17}$$

that is, net worth in previous period plus (minus) profits (losses). Making use of (4.16) and (4.17), it follows that the bankruptcy state occurs whenever:

$$u_{it} < \frac{1}{\phi}\left(g r_{it} - \frac{A_{it-1}}{K_{it}}\right) \equiv \bar{u}_{it} . \tag{4.18}$$

As in Greenwald and Stiglitz (1990, 1993), the probability of bankruptcy ($\Pr^f$) is incorporated directly into the firm's profit function because going bankrupt costs, and such a cost is increasing in the firm's output. Assuming for expositional convenience that u_{it} is uniformly distributed on the support (0,2), and that bankruptcy costs are quadratic, $C^f = cY_{it}^2$ with $c > 0$, the objective function takes the form:[21]

$$\Gamma_{it} = (\phi - g r_{it}) K_{it} - \frac{\phi c}{2}\left(g r_{it} K_{it}^2 - A_{it-1} K_{it}\right) . \tag{4.19}$$

From the first order condition, the optimal capital stock is:

$$K_{it}^d = \frac{\phi - g r_{it}}{c\phi g r_{it}} + \frac{A_{it-1}}{2 g r_{it}} . \tag{4.20}$$

19 It follows that the credit lines periodically extended by the bank to each firm are based on a mortgaged debt contract.

20 As a matter of example, one can think of retooling and adjustment costs to be sustained each time the production process starts.

21 For this program to be well defined g should be such that the condition $A_{it-1} < g r_{it} K_{it}$ holds.

Thus, the desired capital stock in t is decreasing (non-linearly) with the interest rate and it increases linearly with financial robustness, as proxied by the $t-1$ net worth. Time period t desired investment is simply the difference between the desired capital stock and the capital stock inherited from the previous period, $I_{it} = K_{it}^{d} - K_{it-1}$. To finance it, the i-th firm recurs to retained profits and, if needed, to new mortgaged debt, $I_{it} = \pi_{it-1} + \Delta L_{it}$,[22] where $\Delta L_{it} = L_{it} - L_{it-1}$. Making use of (4), the demand for credit is given by:

$$L_{it}^{d} = \frac{(\phi - gr_{it})}{c\phi gr_{it}} - \pi_{it-1} + \left(\frac{1-2gr_{it}}{2gr_{it}}\right) A_{it-1}\,. \tag{4.21}$$

4.5.2 The Banking Sector

We model the banking sector in terms of the reduced form from the working of an oligopolistic industry. The balance sheet of the banking sector is $L_t^s = E_t + D_t$, with L_t being total credit supply, E_t the banks' equity base and D_t deposits which, in this framework, are determined as a residual. To determine the aggregate level of credit supply, we assume that banks are subject to a prudential rule set up by a regulatory body such that $L_t^s = E_{t-1}/\nu$, where the risk coefficient ν is constant. Hence, the healthier are banks from a financial viewpoint, the higher is the aggregate credit supply (Hubbard *et al.*, 2002).

Credit is allotted to each individual firm i on the basis of the mortgage it offers, which is proportional to its size, and to the amount of cash available to serve debt[23] according to the rule:

$$L_{it}^{s} = \lambda L_s \frac{K_{it-1}}{K_{t-1}} + (1-\lambda)\, L_s \frac{A_{it-1}}{A_{t-1}} \tag{4.22}$$

with $K_{t-1} = \sum_{i=1}^{N_{t-1}} K_{it-1}$, $A_{t-1} = \sum_{i=1}^{N_{t-1}} A_{it-1}$, and $0 < \lambda < 1$. The equilibrium interest rate for the i-th firm is determined as credit demand (4.21) equals credit supply (4.22), that is:

$$r_{it} = \frac{2 + A_{it-1}}{2cg\left(\frac{1}{\phi c} + \pi_{it-1} + A_{it-1}\right) + 2cgL_s\left[\lambda\kappa_{it-1} + (1-\lambda)\,\alpha_{it-1}\right]}\,, \tag{4.23}$$

---

22 A word of caution is in order here. The law of motion of the net worth (4.17) seems to imply that the correct time at which profit had to be taken into account in deriving the demand for credit should be time period t. In fact, the timing structure of the model is such that when deciding how much to borrow from banks, firms do not have received any time t revenues yet. Hence, at the beginning of time period t the only internal finance they can count on are inherited equity and time $t-1$ profits.

23 For evidence on the effects exerted by firms' size and credit worthiness on banks' loan policies see e.g. Kroszner and Strahan (1999).

where κ_{it-1} and α_{it-1} are the ratios of individual to total capital and net worth, respectively.

Under the assumption that the returns on the banks' equity are given by the average of lending interest rates $\bar{r}_t$, while deposits are remunerated with the borrowing rate r_t^A, the banks' profit (π_t^B) is given by:

$$\pi_t^B = \sum_{i \in N_t} r_{it} L_{it}^s - \bar{r}_t \left[(1-\omega) D_{t-1} + E_{t-1} \right] \tag{4.24}$$

with $\frac{1}{1-\omega}$ being the spread between lending and borrowing rates. Note that ω, which in what follows will be treated parametrically, captures the degree of competition in the banking sector: the higher is ω, the higher is the interests' spread which, in turn, increases with a higher monopolistic power of banks.

When a firm goes bankrupt, $K_{it} < L_{it}$. In this case, the banking sector as a whole registers a loss equal to the difference between the total amount of credit supplied up to time period t and the relative mortgage, $B_{it} = L_{it} - K_{it} = -A_{it}$, where $A_{it} < 0$ if firm i belong to the set of bankrupt firms Ω_t. Let us call B_{it} *bad debt*. The banking sector's equity base evolves according to the law of motion:

$$E_t = \pi_t^B + E_{t-1} - \sum_{i \in \Omega_{t-1}} B_{t-1} \,. \tag{4.25}$$

Through the banking sector's equity base law of motion, idiosyncratic real disturbances leading to a bankruptcy have systemic consequences: an increase of bad debt forces the aggregate credit supply shifting to the left, thus raising the financial costs due to a higher interest rate, *ceteris paribus*. Furthermore, the distribution of firms' net worth influences the average lending interest rate, which in turn affects the bank's profit and, eventually, credit supply. Thus, firms dynamically affect each other through indirect interactions. In particular, interactions are global and independent of any topological space, and they occur through a field variable, which in our case is the banking sector's balance sheet (Aoki, 1996).

Interactions, if strong enough, allow the system to escape from the property of square root scaling for sums of *iid* shocks due to the Central Limit Theorem. It is well known from statistic theory (e.g, Resnik, 1987) that as N grows large, independence of idiosyncratic disturbances implies that the volatility of the system decays with the square root of size, leading to a power law distribution with exponent $\beta = -0.5$. If *distant* agents are sufficiently correlated through interactions, in turn, aggregate volatility decays more slowly, according to a power law with exponent $\beta < -0.5$. The empirical evidence reported in Amaral *et al.* (1997) for companies and in Canning *et al.* (1998) for countries goes precisely in this direction.

4.5.3 Firms' Demography

Recent empirical work has shown that firms entering and exiting markets contribute almost as much to employment and macroeconomic fluctuations as firms continuing their activity (e.g., Davis *et al.*, 1996). Hence, any theory of business fluctuations should pay particular attention to the way entry and exit of firms are modeled.[24]

In our framework, exits are endogenously determined as financially fragile firms go bankrupt, that is as their net worth becomes negative. Besides making the total output to shrink, exits cause the equity of the banking sector – and, in turn, aggregate credit – to go down. As discussed above, this mean field interaction in terms of a *bank effect* (Hubbard *et al.*, 2002) amplifies and propagates idiosyncratic shocks all over the economy.

As regards entries, the literature has suggested models ranging from exogenously defined purely stochastic processes (Winter *et al.*, 1997), to models where entry is endogenous in that the number of entrants depends on expected profit opportunities (Hopenhayn, 1992). Alas, the available evidence has been so far inconclusive. Caves (1998), for instance, claims that the only firm points are that entrants are in general largely unsure about the probability of prospective success, and that entries does not occur at a unique sector-specific optimal size.

Our modeling strategy aims at capturing these facts by means of a mechanism in which a probabilistic process is affected by prospective performance, and entries can take place at different sizes. First, the number of new entrants (N_t^{entry}) is obtained by multiplying a constant $\bar{N} > 1$ to a probability which depends negatively on the average lending interest rate:

$$N_t^{\text{entry}} = \bar{N}\,\Pr(\text{entry}) = \frac{\bar{N}}{1+\exp\left[d\left(\bar{r}_{it-1}-e\right)\right]}\,, \tag{4.26}$$

where d and e are constants. The higher is the interest rate, the higher are financial commitments, and the lower are expected profits, with entries being lower in number. Second, entrants' size in terms of their capital stock is drawn from a uniform distribution centered around the mode of the size distribution of incumbent firms, each entrant being endowed with an equity ratio ($a_{it} = A_{it}/K_{it}$) equal to the mode of the equity base distribution of incumbents.

4.5.4 Long-Run Dynamics

In order to understand the long-run – i.e., growth – properties of our economy, it is convenient to consider a deterministic version of the model. Indeed, abstracting from uncertainty means getting rid of heterogeneity, so that we

---

[24] Delli Gatti *et al.* (2003) provide an extensive analysis on the relationship between entries and exits and aggregate fluctuations in a model very similar to this one.

can easily keep track of the dynamic behavior of a representative firm. If the interest rate is assumed constant, from (4.16), (4.17) and (4.20) it turns out that the law of motion of the net worth is:

$$A_t = \left(1 + \frac{\phi - gr}{2gr}\right) A_{t-1} + \frac{(\phi - gr)^2}{c\phi gr} \, . \tag{4.27}$$

The solution of this first order difference equation returns the steady state gross growth rate of the economy, $1/2[\phi/gr+1]$, which implies positive growth whenever $(\phi - gr) > 0$: whenever the return to capital is higher than its cost, the economy is characterized by endogenous growth. This result is far from surprising as soon as we note that in our model the production function exhibits constant returns to the only input that can be accumulated, which is the same engine of growth as in the well-known AK endogenous growth model developed by Rebelo (1991).

This analogy can be further extended to appreciate the special role played by credit in our economy. First, recall that in the Rebelo's model the steady-state growth rate depends positively on the saving rate. In our partial equilibrium analysis savings are implicitly defined as the difference between investment and retained profits,[25] so that at each time period t total savings are equal to banks' loans. Indeed, changes in the banking regulatory regime or in the competitive pressure in the banking sector end up affecting the equilibrium lending interest rate, and through it the long-run growth rate.

4.6 Simulation Results: Preliminaries

The complexity of the model directs the analysis of its high-frequency properties towards computer simulation techniques. Figures 4.5 and 4.6 exhibit the evolution of an artificial economy lasting 1000 time periods, implemented using the framework analyzed in the previous section with a starting number of 10,000 firms.[26] In particular, in Fig. 4.5 we show the time path of the logarithm of total output, whereas in Fig. 4.6 it is drawn its volatility expressed in terms of output's growth rates.

From Fig. 4.5 it emerges that our stochastic economy, buffeted with *iid* idiosyncratic disturbances only, is characterized by sizable aggregate fluctuations; that its growth process displays a broken-trend behavior[27] (Perron,

[25] See the *equilibrium* scenario depicted in note 18.

[26] In the following, every simulation were conducted by means of *Swarm*, an agent-based software developed at the Santa Fe Institute to implement artificial economies. Interested readers can find it at the web site: www.swarm.org.

[27] For instance, the average growth rate goes from 0.19% in periods 150–350, to 0.25% in periods 351–780, to 0.37% in periods 780–855, to 0.31% in periods 880–1000. Yearly average growth rates more close to reality could be obtained in this model through a more careful calibration exercise. Given that our main interest is in business fluctuations, however, we leave this undertaking to future research.

1989); and that *Great Depressions* (e.g., the one during the simulation time period 855–880) can suddenly punctuate its time path, due to bankruptcies of great firms that origin remarkable impacts on the business cycle *via* the financial sector (Gabaix, 2005). The output series possesses an autocorrelation parameter equal to 0.99. Interestingly, before large downturns our model seems to exhibit a common pattern: starting from a constant growth trend, the economy gains momentum with accelerating growth and increasing volatility, to subsequently move into a deep recession.

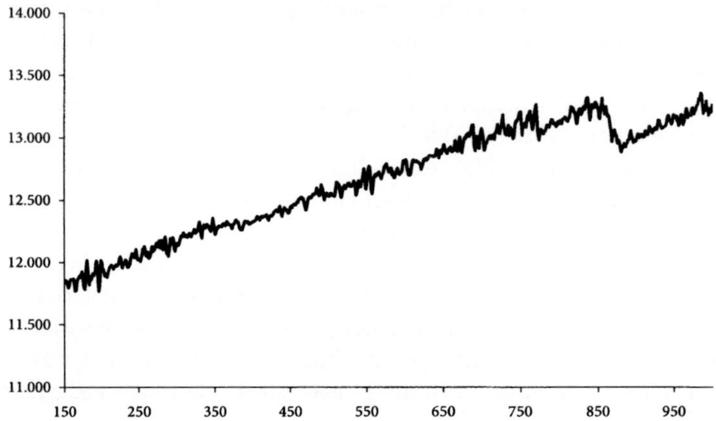

Fig. 4.5. Logarithm of the aggregate output. The first 150 periods have been deleted to get rid of transients

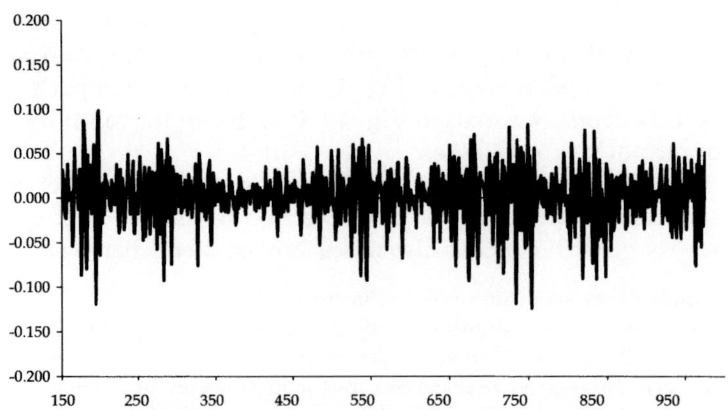

Fig. 4.6. Growth rates of aggregate output

Furthermore, as shown in Fig. 4.6, fluctuations as measured by output's growth rates are characterized by cluster volatility, a well known phenomenon mostly in the financial market literature due to the heavy tails character of asset returns' distributions (Cont *et al.*, 1997). The growth rates' standard deviation is 0.0289.

4.6.1 Firms' Size and Growth Rates Distributions

In Fig. 4.7 we report the Zipf plot for firm sizes recorded at simulation time period 1000. In agreement with recent empirical results (Axtell, 2001) the firms' size distribution is skewed and it follows a power law. Furthermore, the scaling exponents recorded ($\alpha = 1.15$) are consistent with what found in real data (Gaffeo *et al.*, 2003). As widely shown in the complexity literature, the emergence of such a distribution is deeply correlated with the hypothesis of interaction of heterogeneous agents that is at the root of the model. More specifically, the interaction among units buffeted with multiplicative *iid* shocks leads the system's dynamics to a complex critical state in which no attractive point or state emerges. In terms of business fluctuations, it means that there is not a single and determinate equilibrium, but a non-stable state emerges after each recessive or expansive episode.

The firms' size distribution tends to shift to the right during growing phases, while during recessions the estimated stability parameter α decreases. In fact, during expansions greater firms tend to grow faster than smaller ones, causing a higher slope of the interpolating line if compared with the situation observed during recessions. On the contrary, bankruptcies of great firms during downturns cause a more equal distribution of the size distribution. Once again, this is precisely what observed in real data (Gaffeo *et al.*, 2003).

Stanley *et al.* (1996) and Bottazzi and Secchi (2003), among others, find that the growth rates of firms are generally well fitted by a Laplace (or

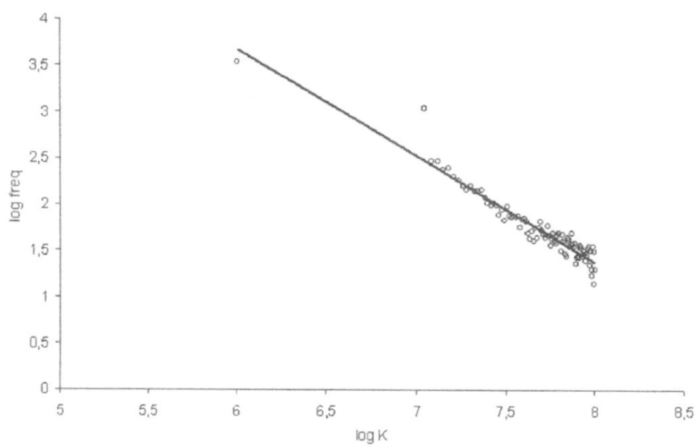

Fig. 4.7. Zipf plot of firm sizes

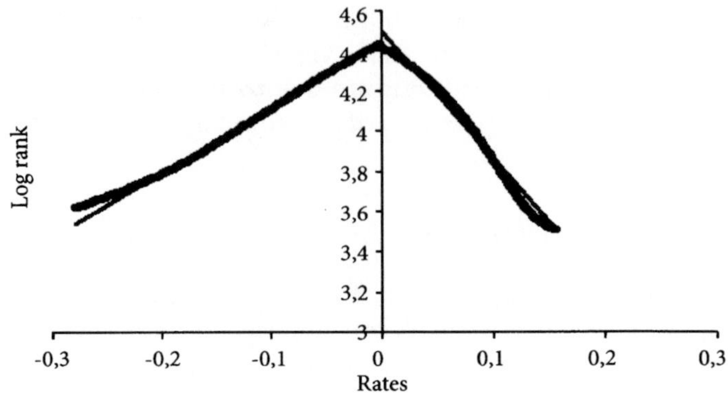

Fig. 4.8. Distribution of firms growth rates

double exponential) distribution. As discussed in Sect. 4.4, such a finding can be shown to derive from firms' size being distributed as a power law. In fact, simulated data for firms' growth rates, reported in Fig. 4.8, are well approximated by a (asymmetric) Laplace distribution.

In another stimulating paper, Lee *et al.* (1998) show that binned growth rates for firms and countries' GDPs settle on the same regression line in a log-log plot. If analyzed from a complex perspective, this result signals the presence of *self similarity*[28], i.e. the behavior of greatest units (countries) reproduces the behavior of smaller units (firms), possibly corrected by a scale factor (Durlauf, 2003). As shown in Fig. 4.9, where we plot the distribution of the growth rates of aggregate output, this feature has been recorded in our model as well. The difference of parameters between the firms' growth rate and the aggregate output growth rates distributions is sensible, in our simulations, to the modeling choice for the production function, though we do not have at this stage any analytical result to prove it.

The model is capable to display several other striking similarities with observable facts. In particular: 1) the frequency of firms' exits seems to be well approximated by an exponential function of firms' age (Steindl, 1965; Fujiwara, 2003); 2) bad debt, that is the amount of unpaid loans due to bankruptcies extended by the banking sector, follows a stretched exponential distribution (Delli Gatti *et al.*, 2003); 3) profits are power law distributed, and exhibit time reversal symmetry (Fujiwara, 2003); 4) expansions and recessions, measured as trough-to-peak and peak-to-trough of the GDP growth rates time series, are distributed as a Weibull (Di Guilmi *et al.*, 2003); 5) the rate of return on the capital (π_i/K_i) and the equity ratio a_i are positively correlated; 6) a higher equity ratio is associated with a lower volatility of

[28] According to Sornette (2000, p. 94), self-similarity occurs when "... arbitrary sub-parts are statistically similar to the whole, provided a suitable magnification is performed along all directions".

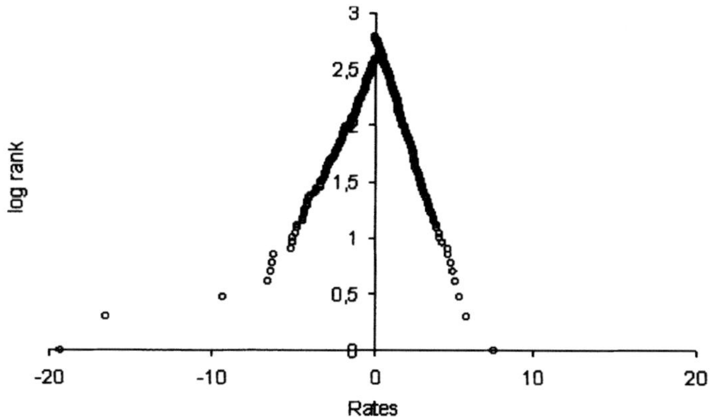

Fig. 4.9. Growth rates of aggregate output

profits, the last two facts being consistent with the evidence one can obtain by analyzing real data.

4.6.2 Conditional Distributions

In this subsection we address a typical aggregation issue, known as the *mixture* problem, which is likely to negatively affect the reliability of results as soon as scaling plots are taken into account (Brock, 1999; Durlauf, 2003). Roughly speaking, the mixture problem asserts that, when aggregating economic units with different behaviors, it is possible to observe marginal distributions with heavy tails even though conditional distributions do not possess such a property. In other terms, a power law may appear simply because heterogeneous units governed by different stochastic processes are erroneously mixed, instead of signaling the invariant properties of a unique Lévy-stable underlying stochastic process. In fact, if the latter is the case one should observe the same scaling behavior independently of which conditioned subsample is considered. In fact, the mixture problem may be present in our work, since the model described in Sect. 4.5 implies different behaviors according to the financial position of firms, as well as differently aged firms.

To understand which variable is likely to be most suitable for conditioning, we start considering one of the basic hypothesis at the root of the model mechanics, that is the fact that an heavy indebted firm is forced to use a large amount of its revenues to pay for its financial commitments, instead of using it for real investments. In other terms, a high leverage is likely to reduce the profitability index. The analysis of the relationship between profit rate and equity ratio, conducted by means of nonparametric regression[29], returns an upward sloping trend (Fig. 4.10) as one would expect from the theoretical model, and in line with what recorded for empirical data. Furthermore,

[29] We use a kernel density estimation (Härdle, 1990), with a Gaussian kernel.

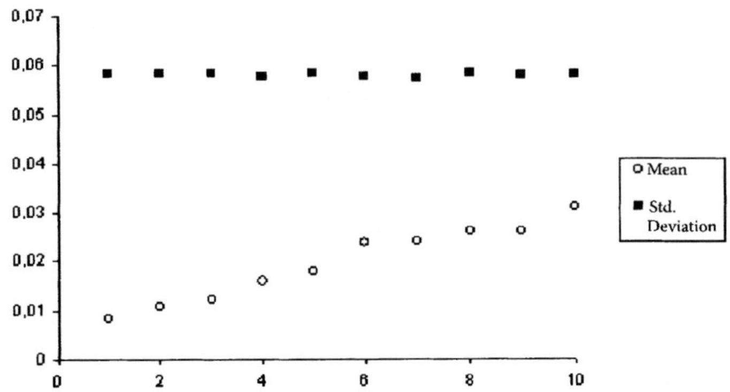

Fig. 4.10. Profit distribution conditioned on the equity ratio, with firms grouped in 10 bins

simulations show that the rate of profit distribution shifts to the right when conditioning on the equity ratio[30], and that the probability to fail does not depend on the size but only on the financial position[31]. This is important for the analysis to follow, suggesting that to address the mixture problem it is sufficient to compute firms' distributions conditional on the equity ratio.

Hence, the power law behavior of the firms' size distribution is analyzed partitioning the $[0, 1]$ interval, to which a belongs, in several bins (chosen according to a percentile allotment). Figure 4.11 shows that data from different distributions conditioned on the equity partition $[0, 0.1734]$, $(0.1734, 0.2269]$, $(0.2269, 0.3319]$[32] collapse on the same interpolating line, a clear sign of self-similarity, thus signaling that the unconditional distribution of firms' size is likely to display a scaling behavior because of its true nature and not due to spurious mixing. The level of the equity ratio seems not to have any influence on the relative growth of firms, since the conditional distributions of growth rates, sorted in bins according to their financial position, invariably collapse on the same curve (Fig. 4.12).

To summarize, of the two forms of heterogeneity in the model – i.e., firms' financial position and age – the one that really matters in firms' behavior is the former, here measured by the equity ratio a. The analysis above shows

[30] The shape of the conditioned profit distributions depends on the assumption one makes on the distribution of idiosyncratic real shocks. We made several trials, to conclude that the best approximation to what observed in real data could be obtained by forcing the relative price shocks to be normally distributed. Nevertheless, none of the model's properties discussed in the main text are qualitatively affected by this modeling choice.

[31] This finding could be further conceived by recalling that firs' exists are exponentially distributed, and that the exponential distribution possesses the well known property of being memoryless.

[32] A fourth bin, i.e. the partition $(0.3319, 1]$, has been excluded from the plot due to a lack of sufficient observations.

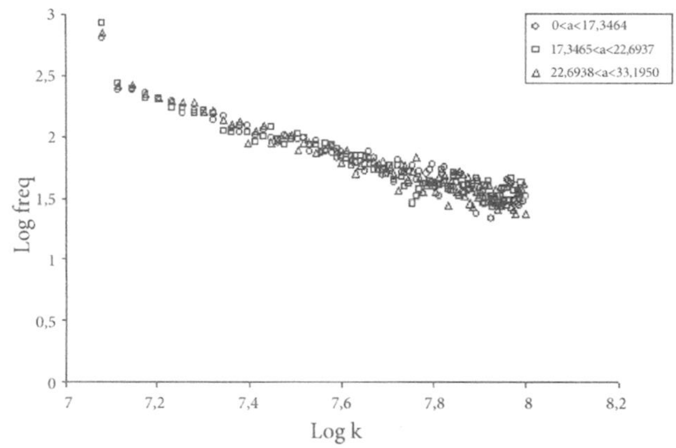

Fig. 4.11. Zipf plot of firms dimension sorted by equity ratio (a)

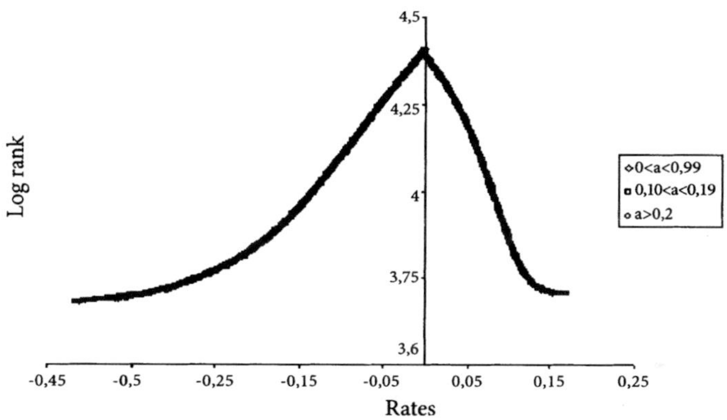

Fig. 4.12. Distributions of firms' growth rates partitioned according to their equity ratio

that the scaling and the self-similarity properties – phenomena which suggest complex behaviors – do not depend on the aggregation of different economic units but it is an intrinsic property of an economic system with interacting units buffeted with idiosyncratic multiplicative shocks.

4.7 Statistical Aggregation

Roughly speaking, the job of a macroeconomist consists in analyzing people's behavior by focusing on the resulting aggregate quantities and their relationships. This is exactly the key point of the *aggregation problem*: starting from the *micro-equations* describing/representing the (optimal) choices of the economic units, what can we say about the *macro-equations*? Do they have the same functional form of the micro-equations (what Theil (1954) called the *analogy principle*)? If not, how to derive the macro-theory? Certainly, simulations constitute a useful way round, and probably the only one when the model is filled with non-linearities. It seems interesting, however, to look for analytically closed-form solution to the aggregation problem even for agent-based model.

The modern approach to aggregation is the one proposed e.g. by Kalejian (1980) and Stoker (1984) which aims at exploiting the statistical structure of the model. In particular, the core idea of the *statistical aggregation procedure* consists in looking for relationships between the first moments of the macro-variables, given the micro-theory, instead of requiring that the aggregate model inherits the same functional form of the micro-relationship.

Suppose that the micro-relationship we are interested in has the following form:

$$y_{it} = f_i(x_{it}, u_{it}, \theta_i) \quad i = 1, 2, \ldots, N \quad t = 1, 2, \ldots, T\,, \tag{4.28}$$

where y_{it} is the strategic choice of the generic economic unit i. In the reduced form representation, that choice depends on the vector of explanatory variables x_{it}, whereas u_{it} is a vector of unobservable characteristics and θ_i a vector of parameters. Call $P(x_{it}, u_{it}, \theta_i; \phi_t)$ the joint distribution of all variables characterized by a time-variant vector ϕ_t. Then, using for the sake of simplicity a continuous notation:

$$E[y_t] = \mu_y(t) = \Psi_y(\phi_t) = \int f(x_t, u_t, \theta_i) P(x_t, u_t, \theta_i; \phi_t)\, \mathrm{d}x_t\, \mathrm{d}u_t\, \mathrm{d}\theta\,, \tag{4.29}$$

while the following equation gives the expected (aggregate) variables:

$$E[x_t] = \mu_x(t) = \Psi_x(\phi_t) = \int x_t P(x_t, u_t, \theta_i; \phi_t)\, \mathrm{d}x_t\, \mathrm{d}u_t\, \mathrm{d}\theta\,. \tag{4.30}$$

The idea behind this line of research is to partition ϕ into two sub-vectors, $\phi_t = (\phi_{1t}, \phi_{2t})$, such that: a) ϕ_{2t} has the same dimension of x_{it}; and b) an invertible function between ϕ_{2t} and $\mu_x(t)$ exists, say $\phi_{2t} = \Psi_x^{-1}(\phi_{1t}, \mu_x(t))$. Conditions a) and b), if combined, imply that we can write:

$$\mu_y(t) = \Psi_y[\phi_{1t}, \Psi_x^{-1}(\phi_{1t}, \mu_x(t))] = F(\mu_x(t), \phi_{1t})\,, \tag{4.31}$$

which represents the (aggregate) relation between the first moments of the economic variables.

Two remarks are in order. First, notice that the aggregate relationship between $\mu_y(t)$ and $\mu_x(t)$ depends on a set of parameters of the joint distribution, here represented by ϕ_{1t}. Second, in general the aggregate relationship F differs from the true micro-one f. In other words, if one allows for heterogeneity in individual choices, it is impossible to use the analogy principle even if we are interested in an aggregate model that holds only on average.

It must be admitted that this approach has been rarely used in the literature, basically because of its difficult implementability and the need for gathering too much information about cross-sectional distributions. Furthermore, as argued by Pesaran (2000), it is far from clear how this approach could be extended to the case of dynamical systems. In what follows, however, we will argue that an approximate and simplified adaptation of the stochastic aggregation procedure can be derived, which possesses the valuable attribute of manageability. Simply stated, the key idea of the aggregation procedure we propose here consists in approximating equations (4.29) and (4.30) along the line suggested by Keller (1980), and extending such a methodology to a dynamical context. This implies to expand around the mean the micro-relation (4.28) in Taylor series up to a certain order k (usually 2), and to take the expectation operator with respect to the agents' cross-section distribution. Obviously, is necessary the existence of a certain number of moments of the distribution. In cases, such the power law distributions describes in the book, in which second moments may not exist we have to use transformations (e.g. the *log*-transformation) giving the number of moments we need.

For expositional purposes let us suppose that, as it is the case in many economic models, the random components u is additive, that the functions f_i and the parameters θ_i are the same across agents ($f_i = f$; $\theta_i = \theta$), and that f does not depend on θ:

$$y_{it} = f(x_{it}) + u_{it} \, . \tag{4.32}$$

Provided that $f(x)$ has the first k+1 derivatives, the deterministic part of (4.32) can be expanded in Taylor series up to order k. In the following, we limit ourselves to the case $k = 2$, but the reader can easily generalize the idea to higher values. In our case, we get:

$$f(x_{it}) = f(\mu_{xt}+\varepsilon_{it}) = f(\mu_{xt})+f'(\mu_{xt})\varepsilon_{it}+\frac{1}{2}\varepsilon_{it}^T f''(\mu_{xt})\varepsilon_{it}+o(||\varepsilon_{it}||) \, , \tag{4.33}$$

where μ_{xt} is the expected value of the vector x_{it}, ε_{it} is the vector spreads from the mean and f' and f'' are, respectively, the gradient and the Hessian matrix of $f(x)$.

Taking the expectation we obtain:

$$E[y_{it}] = f(\mu_{xt}) + \frac{1}{2}E[tr(\varepsilon_{it}\varepsilon_{it}^T f''(\mu_{xt}))] + E[o(||\varepsilon_{it}||)] \, , \tag{4.34}$$

where $E[.]$ is the expectation operator and tr is the trace operator. Using the linearity property of E, we obtain:

$$\mu_{yt} = E[y_{it}] = f(\mu_{xt}) + \frac{1}{2}tr(\Sigma_t f''(\mu_{xt})) + E[o(||\varepsilon_{it}||)] \tag{4.35}$$

where Σ_t is the variance–covariance matrix of x. The equation (4.35) represents the exact aggregate relationship between y and x. When there is no dispersion among agents (i.e., Σ_t converges to the null matrix), (4.35) reduces to the representative agent equation in which the *analogy principle* between first moments of cross-sectional individual behaviors and aggregate variables holds true. In other terms, we can interpret the two terms

$$\frac{1}{2}tr(\Sigma_t f''(\mu_{xt})) + E[o(||\varepsilon_{it}||)]$$

as the error we made using the RAH when in fact there is important heterogeneity among agents/firms[33].

When we try to apply the general exact aggregation procedure discussed so far to real-data problems, we have to distinguish between two cases. First, suppose we have prior information about the agents' distribution. The aggregative relationship between first moments and the parameters of the distribution can be now easily inverted. Consider an individual-level scalar micro-equation (i.e., x is a scalar variable). In this case, (4.35) reads:[34]

$$\mu_{yt} = f(\mu_{xt}) + \frac{1}{2}f''(\mu_{xt})\sigma^2_{xt} + E[o(||\varepsilon_{it}||)] \,. \tag{4.36}$$

Suppose, as a matter of example, that the sample distribution of x can be approximated by an exponential probability density function with parameter b, i.e. $1/b\exp(-x/b)$. Then, the mean μ is equal to b and the variance σ^2 to b^2. This implies that $\sigma^2 = \mu^2$ and the exact aggregate equation becomes:

$$\mu_{yt} = f(\mu_{xt}) + \frac{1}{2}f''(\mu_{xt})\mu^2_{xt} + E[o(||\varepsilon_{it}||)] = h_1(\mu_{xt}) + E[o(||\varepsilon_{it}||)] \,. \tag{4.37}$$

Expanding the micro-relationship in Taylor series of order $k = 2$ implies that the approximate equation

$$\mu_{yt} \approx h_1(\mu_{xt}) \tag{4.38}$$

can be seen as a *second order approximation* (or, more generally, an approximation of order k) to the exact aggregation relation.

If, on the contrary, we do not have prior distributional information, we can usefully exploit any empirical or theoretical information we have about the dynamic evolution of x.[35] For example, let the law of motion of the

[33] In probability theory, those terms represent the error made linearizing a non-linear transformation of random variables. In fact, as discussed in the main text, there is no aggregation error in an economic theory only when the relationships between micro-variables are all linear or affine.

[34] This method may in principle be applied even when second moments of x do not exist (but the first moment of y exists) using an appropriate transformation of the microvariable x. In such cases, the approach can give only qualitative insights.

[35] For instance, in the industrial dynamics literature such a kind of information comes naturally as one considers the evolution of firm sizes. The asymptotic distribution for the size of firms usually associated to alternative models of firms' size growth is log-normal or power law (Axtell, 2001; Sutton, 1997). As it is well known, power law distribution may do not possess second moments. In such situations, we need to work on transformed variables (e.g., logarithmic transformations) to use the above approach. Obviously results must be read carefully since they may hold only qualitatively.

micro-variable x be described by the following first order difference equation:

$$x_{it} = g(x_{it-1}) + z_{it} \ , \tag{4.39}$$

where g is a function of the variable x, and z_{it} is some idiosyncratic component with zero mean and variance δ^2. Assume, as before, that g can be differentiated at least three times, and take an expansion in Taylor's series of order 2 of the right hand side of (4.39). The expected value across all x_{it}, can then be written as an approximate function of the cross-section mean and variance at time $t-1$:

$$\mu_{x,t} \cong h_x(\mu_{x,t-1}, \sigma^2_{x,t-1}) \ . \tag{4.40}$$

Let h_y be the function relating the first moment of y with the first and second moment of x:

$$\mu_{y,t} \cong h_y(\mu_{x,t}, \sigma^2_{x,t}) \ . \tag{4.41}$$

As we take the variance on both sides of (4.39) after the expansion in Taylor's series has been accomplished, it is possible to compute the approximate relation between second moments at time t and first and second moments at time $t-1$

$$\sigma^2_{xt} \approx g'(\mu_{xt-1})^2 \sigma^2_{xt-1} + \delta^2 = v(\mu_{xt-1}, \sigma^2_{xt-1}, \delta^2) \ , \tag{4.42}$$

that can be substituted in the (4.41) to get:

$$\begin{aligned} \mu_{y,t} \cong h_y(\mu_{x,t}, \sigma^2_{x,t}) &= h_y(\mu_{x,t}, v(\mu_{xt-1}, \sigma^2_{xt-1}, \delta^2)) \\ &= h_{yy}(\mu_{x,t}, \mu_{xt-1}, \sigma^2_{xt-1}, \delta^2) \ . \end{aligned} \tag{4.43}$$

Then, by inverting (4.40) with respect to $\sigma^2_{x,t-1}$ we obtain:

$$\sigma^2_{x,t-1} \cong l_x\left(\mu_{x,t}, \mu_{x,t-1}\right) \ , \tag{4.44}$$

and by replacing (4.44) in (4.43) we get a second (approximate) aggregate relationship:

$$\mu_{y,t} \cong h_2(\mu_{x,t}, \mu_{xt-1}, \delta^2) \ , \tag{4.45}$$

relating the per-capita value of the variable y_i at time t with the per-capita value of the variable x_i at time t and $t-1$, and with the variance of the idiosyncratic shock affecting the dynamics of x.

Summarizing, whenever we are interested in describing the aggregate implications of a stochastic process x characterizing the evolution of a population of agents, we suggest to approximate the dynamic evolution of the first two moments (in general the first k moments) by the following map:

$$\begin{cases} \mu_{xt+1} \approx g(\mu_{xt}) + \dfrac{1}{2} g''(\mu_{xt}) \sigma^2_{xt} \\ \sigma^2_{xt+1} \approx g'(\mu_{xt})^2 \sigma^2_{xt} + \delta^2 \end{cases} . \tag{4.46}$$

In Fig. 4.13 and 4.14, we present the results of a simulation of the equity ratio stochastic process for a simplified version of the model described in Sect. 4.5, in which the interest rate has been considered constant. A set of 10,000 firms has been simulated for 20 periods. At the start, all the firms share the same equity ratio, set at a value of 70%, and an initial variance equal to 0. For each period, the average and the variance of the firm equity ratio distribution has been recorded, and plotted as open circles. In both figures, the solid line represents the approximation given by system (4.46).

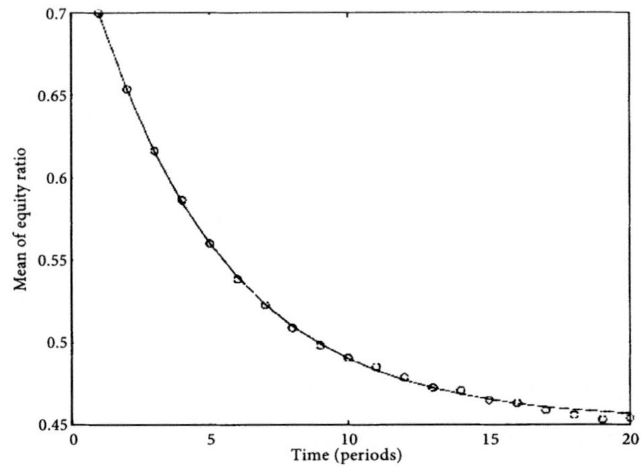

Fig. 4.13. Mean of equity ratio. The dots represent the results of the stochastic simulation, whereas the solid line is the solution of the system (4.46)

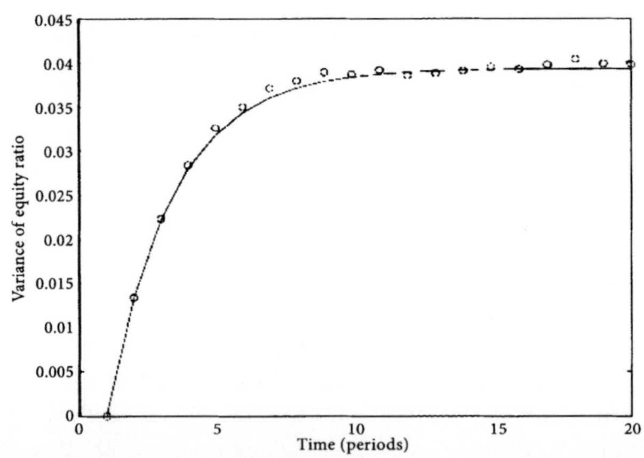

Fig. 4.14. Variance of equity ratio. The dots represent the results of the stochastic simulation, whereas the solid line is the solution of the system (4.46)

4.8 Summing Up

Scaling phenomena and power law distributions are rather unfamiliar concepts for scholars interested in business cycle theory, regardless of the fact that these objects have been studied in economics since a long time. The reason for this neglectfulness should be looked for in the reductionism methodology which has so far permeated modern macroeconomics. Our position is that the reductionism paradigm is not only theoretically unsatisfying, but it can also be falsified as soon as proper new stylized facts are isolated. Concepts and methods inspired from physics have revealed particularly useful in detecting new facts and guiding theory formation. This work aims at popularize the scaling approach to business fluctuations, by discussing some scaling-based ideas involved in viewing the macroeconomy as a complex system composed of a large number of heterogeneous interacting agents (HIAs).

In particular, in this chapter we present a simple agent-based model of the *levered aggregate supply* class developed by Greenwald and Stiglitz (1990, 1993), whose core is the interaction of heterogeneous financially fragile firms and a banking sector. In order to grasp the empirical evidence we adopt a methodological approach based on agent-based simulations of a system with HIAs. In our framework, the origin of business fluctuations can be traced back to the ever changing configuration of the network of heterogeneous interacting firms.

Simulations of the model replicate, with a good level of approximation, a set of stylized facts, particularly two well known *universal laws*: i) the distribution of firms' size (measured by the capital stock) is skewed and described by a power law; ii) the distribution of the rates of change of aggregate and firms' output shows a fat tail double exponential (Laplace) shape. So far, the literature has dealt with stylized facts (i) and (ii) as if they were unrelated. We have discussed as that power law distribution of firms' size may lays at the root of the fat tails distribution of growth rates.

The model can be extended in a number of ways to take into account, among other things, the role of aggregate demand, different degrees of market power on the goods and credit markets, technological change, policy variables, learning processes, etc. Our conjecture, however, is that the empirical validation of more complex models will be due to the basic ingredients already present in the benchmark framework: the power law like distribution of firms' size and then associated double exponential like distribution of growth rates which in turn can be traced back to the changing financial conditions of firms and banks.

Appendix

In this appendix we briefly describe the assumptions and procedures we followed to simulate the model. A simulation is completely described by the parameter values, the initial conditions and the rules to be iterated period after period. First of all, we set the parameter values and the initial conditions for state variables needed to start the simulation. There parameters of the model are relative to the firm, bank and the entry process.

For the firm we have:

- the productivity of capital ϕ,
- the parameter of the bankruptcy cost equation c,
- the firm's equity-loan ratio α,
- the variable cost parameter g

For the bank:

- the mark down on interest rate ω,
- the weight the bank gives to the capital in allotting the credit supply λ.

For the entry process:

- the location parameter d,
- the sensitivity parameter e,
- the size parameter $\bar{N}$.

They are set as follows: $\phi = 0.1$; $c = 1$; $\alpha = 0.08$, $g = 1.1$; $\omega = 0.002$, $\lambda = 0.3$, $d = 100$, $e = 0.1$. $\bar{N}$ must be set according to the initial number of firms (see below). The first step of the simulation occurs at time $t = 1$. To perform calculations in period 1 for each firm we must set initial conditions for firms' capital, the equity base, profit and bad debt. We chose the following values $K_{i0} = 100$, $A_{i0} = 20$, $L_{i0} = 80$, $\pi_{i0} = 0$, $B_{i0} = 0$. We run simulations for several values of initial firms and for different number of iterations. In the simulation we report on Sect. 4.6 the initial number of firms was set to 10,000 and the number of iterations to 1000. Given the initial number of firms, we set $\bar{N} = 180$. The main loop is described in the following algorithm.

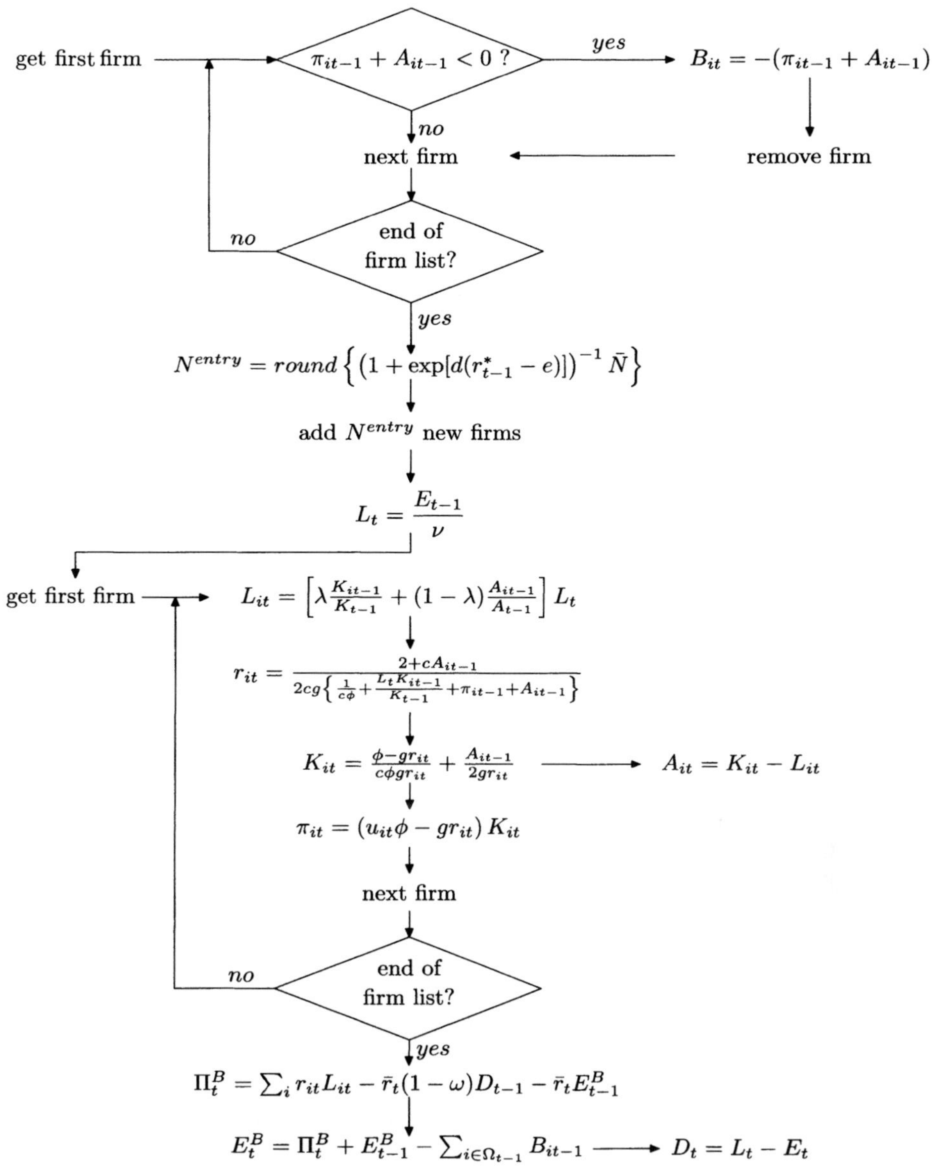
get first firm
$\pi_{it-1} + A_{it-1} < 0$?
yes
$B_{it} = -(\pi_{it-1} + A_{it-1})$
no
next firm
remove firm
no
end of firm list?
yes
$N^{entry} = round\left\{\left(1 + \exp[d(r^*_{t-1} - e)]\right)^{-1} \bar{N}\right\}$
add N^{entry} new firms
$L_t = \frac{E_{t-1}}{\nu}$
get first firm
$L_{it} = \left[\lambda \frac{K_{it-1}}{K_{t-1}} + (1-\lambda)\frac{A_{it-1}}{A_{t-1}}\right] L_t$
$r_{it} = \frac{2 + cA_{it-1}}{2cg\left\{\frac{1}{c\phi} + \frac{L_t K_{it-1}}{K_{t-1}} + \pi_{it-1} + A_{it-1}\right\}}$
$K_{it} = \frac{\phi - g r_{it}}{c\phi g r_{it}} + \frac{A_{it-1}}{2 g r_{it}}$
$A_{it} = K_{it} - L_{it}$
$\pi_{it} = (u_{it}\phi - g r_{it}) K_{it}$
next firm
no
end of firm list?
yes
$\Pi^B_t = \sum_i r_{it} L_{it} - \bar{r}_t (1-\omega) D_{t-1} - \bar{r}_t E^B_{t-1}$
$E^B_t = \Pi^B_t + E^B_{t-1} - \sum_{i \in \Omega_{t-1}} B_{it-1}$
$D_t = L_t - E_t$

5 Where Do We Go from Here?

5.1 Instead of a Conclusion

At the end of this book, it is clear to us that the research carried out so far is only a step in a much longer intellectual voyage. We feel the need to pause and reflect not only on the distance already covered but also on the direction we have to take for future research. For this simple reason, this concluding chapter is not a conclusion at all. We want to open a window on the future to foresee the shifting ground of research in contemporary macroeconomics and position ourselves, our incomplete and as yet probably inadequate set of ideas, methods and tools in the debate that will come.

5.2 Where Are We?

The research on the role of agents' heterogeneity in shaping microeconomic behaviour and macroeconomic performance has been a thriving industry in the profession for the last ten years or so.[1] The representative agent assumption is still the cornerstone of most of contemporary macroeconomics but the awareness of its limitations is spreading well beyond the circle of more or less dissenting economists. Also in mainstream macroeconomics, in fact, the representative agent is not as eagerly embraced as in the early years of the debate on microfoundations in the remote '70s.

In order to take heterogeneity seriously in macroeconomic modelling, one should start with heterogeneous behavioural rules at the micro level and determine the aggregate (macroeconomic) quantity – such as GDP – by adding up the levels of a myriad of individual quantities. Statistical regularities at the aggregate level are characterized by *emerging properties* which do not show up at the microscopic level. The evolving macroeconomic features of the economy, in turn, feed back on microscopic behaviour in many ways, for instance by means of externalities or mean field effects.

[1] The literature on issues pertaining to heterogeneity is growing at a very high exponential rate. An early overview of the role of heterogeneity in macro-dynamic models of the '90s in Delli Gatti, Gallegati and Kirman (2000). Hommes (2006) surveys models with heterogeneous agents focusing mainly on financial markets.

The increasing availability of computational power has allowed the implementation of this bottom-up procedure in multi-agent models. Not surprisingly, in the last ten years, the development of Agent Based Modelling (ABM) has impressed a formidable boost to research on issues concerning heterogeneity.[2] Multi-agent modelling is the most straightforward way of tackling the heterogeneity issue. In the profession at large, however, there is no agreement on the opportunity of following this methodology. While some economists, mainly in the unorthodox camp, eagerly embrace the new research strategy, some others, mainly in the mainstream, are skeptical or even dismissal.

There are at least three reasons for this skepticism: (i) a basic distrust for the output of computer simulations, which is generally very sensitive to the choice of initial conditions and parameter values; (ii) a critique of the prevailing research strategy in ABM, whose pillars are adaptive micro-behavioural rules and out-of-equilibrium processes, often considered ad hoc; (iii) the difficulty and sometimes the impossibility of thinking in macroeconomic terms, i.e. of using macro-variables in the theoretical framework.

The first type of skepticism is rapidly fading away. After all, also Real Business Cycle theory – the benchmark line of thought in neoclassical macroeconomics – produces models that are too complicated to be solved by pen and paper and must be simulated. In order to do so RBC theorists have developed procedures to calibrate their models which, with the passing of time and the spreading in the profession, have become standard tools – we can even call them protocols – of macroeconomic research.

As to the behavioural rules at the micro-level, it is true that some of the most enthusiastic believers in the economy as a Complex Adaptive System have seized the opportunity of agent based modelling to propose rules of individual behaviour characterized by bounded rationality in environments characterized by uncertainty, learning and adaptation. In fact, differently from mainstream economics, the ABM approach is particularly suitable to address issues of heterogeneity, interaction and complexity.[3] Multi-agent models allow the comparison of the impact of different behavioural rules of thumb, which are often traced back to bounded rationality and adaptive behaviour. There is no reason, however, to assume that this is the only way of modelling

---

[2] Tesfatsion and Judd (2006) provides a comprehensive survey of the different viewpoints from which the exploration of behaviour of individual agents can be carried out.

[3] The practice of combining heterogeneity and interactions is at odds with mainstream macroeconomics which is unable, by construction, to explain non-normal distributions, scaling behavior or the occurrence of large aggregate fluctuations as a consequence of small idiosyncratic shocks. As Axtell (1999: 41) claims: "given the power law character of actual firms' size distribution, it would seem that equilibrium theories of the firm [. . .] will never be able to grasp this essential empirical regularity."

individual choices. The multi-agent framework can also accommodate models of optimizing behaviour of heterogeneous agents.

It is also true that ABM is the most straightforward way of treating out-of-equilibrium dynamic processes.[4] There is no reason, however, to assume that the dismissal of equilibrium is a necessary ingredient of any AB model. Out-of-equilibrium dynamics is one way of configuring a collection of agents' choices. Maybe it is also the most reasonable or realistic one. The multi-agent framework, however, can also accommodate models of market equilibrium in the presence of heterogeneous agents.

The framework presented in Chap. 4, Sect. 4.5, for instance, is a *hybrid model*. First of all, in that model optimizing behaviour (on the part of heterogeneous firms which are pursuing the optimal degree of capital accumulation) co-exist with non-optimizing behaviour (on the part of the banking system which determines the supply of credit on the basis of a prudential rule of thumb). Second, in the model equilibrium on the credit market coexist with disequilibrium on the goods market.

Finally, the difficulty of thinking in macroeconomic terms can be eased by means of an appropriate aggregation procedure. For instance the stochastic aggregation procedure discussed in Chap. 4, Sect. 4.7, allows to resume macroeconomic thinking in a multi-agent framework.[5] We claim that this aggregation procedure is a feasible alternative to the Representative Agent.

5.3 A Hybrid Framework

In our opinion, the capability of a model to reproduce significant stylized facts both at the micro and at the macro level is a distinctive feature of a good modelling strategy. For instance, as discussed above, empirical data have shown that firms size distribution is approximated by a power law (Axtell, 2001; Gaffeo *et al.*, 2003) and aggregate and firm's growth rates are often approximated by a double exponential (Laplace) distribution (Stanley *et al.*, 1996; Bottazzi and Secchi, 2003).

In this context, since the distribution of firms' size follows, with a certain degree of approximation, a power law we expect idiosyncratic shocks to "big firms" to be responsible to a non negligible extent of the ensuing turning point. In fact small idiosyncratic shocks at firm-level may generate large aggregate fluctuations when firms' size is power law distributed (Gabaix, 2005).

The ambitious aim of the model of Sect. 4.5 consists in reproducing the empirical evidence by means of a macrodynamic framework in which financial factors play a crucial role. For the sake of discussion, let's recapitulate and clarify the modelling strategy we adopted.

---

[4] See Arthur (2006) for a thorough treatment of this point.

[5] For a thorough discussion of the procedure and comparison with other aggregation procedures see Gallegati *et al.* (2006).

Starting from the assumption, well corroborated by the existing evidence, that firms differ from one another according to their financial conditions, captured by the *equity ratio*, we build a macrodynamic model in three steps. First of all we derive a behavioural rule at the microeconomic level for investment activity in an optimizing framework. We have adopted an optimizing perspective precisely to show that a multi-agent framework can accomodate optimizing behavior. Following Greenwald and Stiglitz (1993) each firm is assumed to maximize expected profit less expected bankruptcy costs. From the optimization we derive individual investment, output and demand for credit as a function of the individual equity ratio.

Second, we model the credit market. The aggregate supply of bank loans is determined as a multiple of the aggregate net worth of the banking system. The aggregate demand for loans is obtained by summing up the individual financial needs. It is worth-noting that we do impose an equilibrium condition at the aggregate level for the credit market. Moreover we assume a simple rule for the allotment of aggregate credit to firms: in equilibrium, the allocation of credit to each and every borrower is determined by the availability of collateralizable assets. The single most important outcome of the allocation of credit in equilibrium is the determination of the interest rate charged to each and every borrower, which ultimately depends – among other things – by the net worth of the borrower itself and of the banking system.

The net worth of the banking system is the crucial vehicle of (indirect) interaction into microeconomic behaviour. For instance, a positive feedback mechanism is activated by the bankruptcy of a single firm due to the negative impact that the non-performing loan (or bad debt) of the bankrupt firm has on the net worth of the banking system. The aggregate supply of credit shrinks, being a multiple of the bank's net worth, and the interest rates go up for each and every borrower. As a consequence of the interest rate hike, the most fragile firms may be forced into bankruptcy. In other words *indirect interaction among firms* through the credit market may force vulnerable firms to exit the market. This fact may start a *chain reaction* and lead to an avalanche of bankruptcies.

In order to endogenize the dynamics of the distribution, we focus on the law of motion of the individual firm's equity which is a function, among other things, of the interest rate. The third step consists in plugging the equilibrium value of the interest rate into the individual law of motion. Inasmuch as the interest rate depends on the net worth of the banking system, the individual net worth turns out to be a function also of the financial conditions of the other firms. In a sense we incorporate a macrofoundation of the micro-dynamics. The individual laws of motion are simulated in a multi-agent setting and macroeconomic aggregates are determined by adding up individual quantities. The moments are computed directly from the empirical distribution obtained from simulated data.[6]

In order to describe in the simplest form the modeling strategy we have adopted, let's write the microeconomic behavioural rule as follows

$$x_i = \phi(f_i, r_i) \ ,$$

where x_i is a choice or control variable (capital in the model of Sect. 4.5) for the i-th agent, f_i is an indicator of financial robustness (financial condition or position; equity or net worth in our case) of the i-th agent, r_i is en endogenous variable pertaining to the the i-th agent (the interest rate in our case). The microeconomic behaviour can be ad hoc, adaptive or optimizing (as in our case).

In the model of Sect. 4.5 *in equilibrium*, at the firm level the interest rate is determined when the individual financial need matches credit availability for the single borrower. The financial need of the i-th agent $l_i^d = x_i - f_i$ is the difference between the level of the choice variable and the financial position. Credit availability $l_i^s = s(x_i)\mu e$ is a share $s(x_i)$, of aggregate credit supply μe, which in turn is a multiple of the financial position of the banking system e.

Equaling financial need and credit availability, taking into account the behavioural rule above and solving for r_i one gets the equilibrium value of the endogenous variable: $r_i = r(f_i, \mu e)$. Notice however, that the net worth of the banking system depends on aggregate financial condition of the firms $e = e(f)$ hence $r_i = r(f_i, \mu e(f))$. Plugging this expression into the behavioural rule one gets $x_i = \phi(f_i, r(f_i, e(f)))$ or, in simpler form

$$x_i = \xi(f_i, f) \ , \tag{5.1}$$

where $\xi_{fi} = \phi_{fi} + \phi_r r_{fi}$; $\xi_f = \phi_r r_e e_f$.

Equation (5.1) shows the interaction at work through a mean field effect. The individual choice variable depends on the individual financial conditions and on the average or aggregate financial condition. This is the cornerstone of modeling financial-real interrelations in an heterogeneous setting.

Dynamics is introduced through the law of motion of the individual financial condition: $f_i' = g(f_i, x_i, r_i)e' = h(e, f)$ where $'$ is the unit time advancement operators. Plugging the corresponding expressions for x_i and r_i into the laws of motion one gets mean field effects at work over time.

Let's apply the aggregation procedure presented in Sect. 4.7. Take a linear approximation in Taylor's series up to the second order term in $E(f_i) = fx_i \approx \xi(f) + \xi_f(f)(f_i - f) + 1/2\xi_{ff}(f)(f_i - f)^2$.

6 As an alternative, one can apply an aggregation procedure to the individual law of motion and determine a two dimensional non-linear dynamic system in discrete time which describes the evolution over time of the mean and the variance of the distribution itself. For an example see Agliari *et al.* (2000).

Summation and averaging yields

$$x = E(x_i) \approx \xi(f) + \xi_f(f) E(f_i - f) + \frac{1}{2}\xi_{ff}(f) E(f_i - f)^2 \ ,$$

where $E(.)$ is the expectations operator. Notice that by construction $E(f_i - f) = 0$ and $E(f_i - f)^2 = V_f$. Hence

$$x \approx \xi(f) + \frac{1}{2}\xi_{ff}(f) V_f \ .$$

The aggregate or macroeconomic (choice) variable depends not only on the mean but also on the variance (and higher moments) of the distribution of financial conditions. Hence the shape and evolution of the distribution of financial conditions is important for macroeconomic performance.

Equation (5.1) can emerge in many different context. Here is a simple example. Assume the behavioural rule is $x_i = \phi(f_i, r)$. Applying the aggregation procedure presented in Sect. 4.7 one gets

$$x \approx \phi(f, r) + \frac{1}{2}\phi_{ff}(f, r) V_f \ .$$

The demand for credit (in the aggregate) is $l^d = x - f$. The supply is $l^s = \mu e$. When demand and supply are in equilibrium, the interest rate (uniform across firms) is $r = r(f, V_f, e)$. The net worth of the banking system, however, is a function of aggregate financial conditions of firms. Hence $x_i = \xi(f_i, f, V_f)$ which is a variant of (5.1).

5.4 Where Do We Go from Here?

The model presented in the previous chapter and recapitulated in simplified form in the previous section is no more than a first step in the direction we want to follow. Our aim is to produce *generative macroeconomics.*[7] Macroeconomics, in fact, should have microfoundations: we do not agree with a purely holistic approach to macro-modeling. The appropriate microfoundations, however, must take into account heterogeneity and interaction. Moreover microeconomic behaviour should be not be modeled in isolation because it is deeply affected by the macroeconomic scenario. The impact of macroeconomic developments on microscopic choice may be mediated by an equilibrium configuration of one or more markets as shown in the previous section. However, this is not necessarily the rule. Macroeconomic externalities and mean field effects can affect microeconomic behaviour also directly. In other words, a good research strategy is based on an explicit consideration of a two-way causation link between micro-behaviour and macro-variables.

[7] We borrow this expression from the idea of a generative social science. See Arthur (2006)

ABM is a great leap forward in the effort to equip the profession with the appropriate tools to deal with heterogeneity and interaction in macroeconomics. As we have seen above, ABM does not necessarily imply the dismissal of optimization and equilibrium. We can move beyond the borders of our hybrid framework, however in two directions: (i) assumptions can and should be more realistic, i.e. agents' actions and market processes in the model should mimic real ones; (ii) the model should be complete, i.e. consider goods, labour and financial markets and their interrelations.

As to realism, we start from the assumption that agents do their best to survive and possibly gain a satisfying level of consumption (in the case of households) or profit (in the case of firms) in a market environment in which uncertainty is pervasive. In this context, *transaction costs* are relevant and market processes and institutions are designed – or emerge spontaneously – as a way of saving on these costs. Moreover, a certain degree of market power on the part of firms is a necessary part of the picture. The microeconomic behaviour of agents on markets, therefore, can be described as a process of *adaptation* to a difficult market environment in a complex economy. In a nutshell, we must model the way in which production, pricing, capital accumulation and financing occur by means of procurement processes (Tesfatsion and Judd, 2006) and is dictated by the need to carry out *procurement processes*.

The main building blocks of a framework consistent with the modelling assumptions spelled out above are the following. In each period, each firm hires labour and invest to produce consumption goods. The firm basically knows only a limited neighbourhood of the initial condition (the *status quo*) on the "demand curve" for its products. The desired scale of production and sale price, therefore, are constrained by expected demand. *Demand expectations* change over time by means of an adaptive mechanism: expectations – and therefore output and price – are revised upward (downward) if a firm experienced excess demand (supply) in the previous period. The degree of expectation revision maybe stochastic.

The interactions in the goods market should be described by a *search and matching* process between the firm, which sells the final output, and households. Prices (and wages) are posted by firms and discovered by households under a thick veil of ignorance. In this market environment, *the Walrasian auctioneer is conspicuous for his absence.*

The household and the firms are picked at random from a distribution and prices and quantities are discovered in firms' stores. The household sorts the prices and the corresponding stores in ascending order (from the lowest to the highest price) and spends all her wealth in goods of the firms with the lowest price. If a firm has not all the quantity the household wants to buy, she spends the remaining wealth in the firms with the second lowest price and so on. The wealth leftover after a certain number of visits is saved for the future. The entire procedure is repeated until every household with a positive wealth has been drawn. At the end of the buying-selling process, firms have received orders and implemented sales.

Wages and vacancies are posted by firms and discovered by households/workers. The wage rate offered by the firm changes over time by means of an adaptive mechanism: the firm revises the wage rate upward if the search for labour was not successful – that is if not all posted vacancies were filled – in the previous period. It is revised downward in the opposite case. The degree of wage rate revision maybe stochastic. The worker adjusts her reservation wage taking into account both the employment status and price inflation in the previous period.

The matching process which determines employment and wages goes as follows. A a firm and a worker are drawn at random from a discrete distribution. The firm hires the worker if and only if it has still an open vacancy and the worker's reservation wage is less or equal to the wage bid. If the worker is not hired by firm, another firm is drawn and a new iteration starts. If the worker is not hired at end of a certain number of iterations, then he stays unemployed in the period and earns zero income.

In this context, firms are profit seeking agents but not necessarily profit maximizers. The attempt to escape bankruptcy and survive is top ranking in the agenda of the adaptive firm. Therefore the issue of financial vulnerability is crucial over the entire lifecycle of the firm.

The firm is endowed with a certain initial level of net worth. External financing basically coincides with credit extended by banks. Employment and production plans are implemented if the firm has enough funds to finance them. Assuming that all the profits are retained within the firm, the equity base evolves over time according adding realized profits to the net worth inherited from the past. If a firm ends up with a negative net worth, it exits the market and is replaced by a new entrant firm.

Procurement processes on the credit market may be modelled as a search and matching process similar to the ones presented above for the goods and labour markets. Interest rates and bank loans are posted by banks and discovered by firms. The interest rate offered by the bank changes over time by means of an adaptive mechanism: the bank revises the interest rate downward if the search for a borrower was not successful – that is if not all posted bank loans were filled. It is revised upward in the opposite case. The degree of interest rate revision maybe stochastic or depend on the financial condition of the prospective borrower. In this context the process of liquidity creation and distribution itself can be modelled as a procurement process on the interbank market in which banks and the central bank interact.

Alternatively, one can think of a centralized credit market in which banks act as a system with respect to a decentralized and fragmented corporate sector. The interest rate charged by the banking system maybe uniform across firms or differentiated from firm to firm. In both cases, the financial conditions of firms are an important determinant of the interest rate policy of the banking system.

Once empirically corroborated, the *complex adaptive system* described so far can be put to work to derive insights on the working of policy moves. It would be interesting, for instance, to assess the ways in which a change in the interest rate controlled by the central bank affects the interbank market and trickles down to the economy at large, showing up indirectly in sequence in the procurement processes of the financial, goods and labour markets. A different transmission mechanism of policy moves could be envisaged for fiscal policy and industrial policy.

As the reader can realize at the end of the wish list of modelling choices for future research presented in this section, the distance covered so far and succinctly described in the present book, is relatively short. We are on a trajectory in the transitional dynamics towards a new and exciting description of the economy. The destination seems far away and – more important – not absolutely clear in advance: we are still groping towards a more convincing description of the economy we want to explore.

The only consoling consideration we can think of is that we are not ardent believer in the existence of a unique steady state, let alone global stability. Regardless of the precise contours of the final destination and the distance to be covered, moving forward is the only really important thing.

References

1. A. Agliari, L. Gardini, D. Delli Gatti, and M. Gallegati, Global Dynamics in a Nonlinear Model of the Equity Ratio., *Chaos, Solitons & Fractals*, vol. 11, n. 6, pp. 961–985, 2000.
2. P. R. Agenor, J. McDermott, and E. Prasad. Macroeconomic Fluctuations in Developing Countries: Some Stylized Facts. *World Bank Economic Review*, 14:251–285, 2000.
3. L. Amaral, S. Buldyrev, S. Havlin, H. Leschhorn, P. Maas, M. Salinger, E. Stanley, and M. Stanley, Scaling Behavior in Economics: I. Empirical Results for Company Growth, *Journal de Physique*, 7:621–633, 1997.
4. L. Amaral, S. Buldyrev, S. Havlin, M. Salinger, and E. Stanley, Power Law Scaling for a System of Interacting Units with Complex Internal Structure, *Physical Review Letters*, 80:1385–1388, 1998.
5. J. L. Anderson, R. E. Kates, L. S. Kegeles, and R. G. Madonna. Divergent Integrals of Post-Newtonian Gravity: Nonanalytic Terms in the Near-Zone Expansion of a Gravitationally Radiating System Found by Matching. *Phys. Rev. D*, 25(8):2038–2048, Apr 1982.
6. M. Aoki. *New Approaches to Macroeconomic Modeling*, Cambridge University Press, Cambridge, 1996.
7. M. Aoki. A Simple Model of Asymmetrical Business Cycles: Interactive Dynamics of a Large Number of Agents with Discrete Choices. *Macroeconomic Dynamics*, 2:427–442, 1998.
8. H. Aoyama, W. Souma, Y. Nagahara, M. Okazaki, H. Takayasu, and M. Takayasu. Pareto's Law for Income of Individuals and Debt of Bankrupt Companies. *Fractals*, 8:293, 2000.
9. K. Arrow. *Social Choice and Individual Values.* New York, Wiley & Sons, 1951.
10. K. Arrow. Rational Choice Functions and Ordering. *Economica*, 1959.
11. K. Arrow. A Utilitarian Approach to the Concept of Equality in Public Expenditures. *Quartely Journal of Economics*, 1971.
12. W. B. Arthur. *Out-of-Equilibrium Economics and Agent-Based Modeling*, In: L. Tesfatsion, K. L. Judd (eds.), Handbook of Computational Economics, Volume 2: Agent-based Computational Economics, Elsevier Science B.V., to appear, 2006.
13. R. Axtell. The Emergence of Firms in a Population of Agents: Local Increasing Returns to Scale, Unstable Nash Equilibria, and Power Law Size Distributions. Working paper. The Brookings Institution, 1999.
14. R. Axtell. Zipf Distribution of US Firm Sizes. *Science*, 293:1818–1820, 2001.

15. R. Axtell, M. Gallegati, and A. Palestrini, Common Components in Firms' Growth and the Sectors Scaling Puzzle. Working paper available at SSRN: http://ssrn.com/abstract=1016420, 2006.
16. P. Bak. *How Nature Works. The Science of Self-Organized Criticality.* Oxford University Press, Oxford, 1997.
17. P. Bak, K. Chen, J. Scheinkman, and M. Woodford. Aggregate Fluctuations from Independent Sectoral Shocks: Self-Organized Criticality in a Model of Production and Inventory Dynamics, *Ricerche Economiche*, 47:3–30, 1993.
18. E. J. Bartelsman, S. Scarpetta, and J. Halitwanger. *Distributed Analysis of Firm-Level Data from Industrial and Developing Countries.* Mimeo, 2003.
19. W. Baumol. Productivity Growth, Convergence, and Welfare. *American Economic Rewiew*, 76:1072–1085, 1986.
20. R. Bénabou. Heterogeneity, Stratification and Growth: Macroeconomic Implications of Community Structure and School Finance. *American Economic Review*, 86:584–609, 1996.
21. O. J. Blanchard and N. Kyotaki. Monopolistic Competition and the Effects of Aggregate Demand. *American Economic Rewiew*, 77:647–666, September 1987.
22. A. Blank and S. Solomon. Power Laws in Cities Population, Financial Markets and Internet Sites (Scaling in Systems with a Variable Number of Components). *Physica A*, 287:279–288, 2000.
23. G. Bottazzi. A Comment on the Relationship Between Firms' Size and Growth Rate. LEM Working Paper, 2007.
24. G. Bottazzi, E. Cefis, and G. Dosi. Corporate Growth and Industrial Structures: Some Evidence from the Italian Manufacturing Industry. *Industrial and Corporate Change, Oxford University Press*, 11(4):705–723, August 2002.
25. G. Bottazzi and A. Secchi. Why Are Distribution of Firms Growth Rates Tent-Shaped?, Economic Letters, 80, pp. 415–420, 2003.
26. S. Bowles. Endogenous Preferences: The Cultural Consequences of Markets and Other Economic Institutions. *Journal of Economic Literature*, 36:75–111, 1998.
27. S. Brakman, H. Garretsen, C. Van Merrewijk, and M. Van Den Berg. The Return of Zipf: Towards a Further Understanding of the Rank-Size Distribution, *Journal of Regional Science*, vol. 39, pp. 183–213, February 1999.
28. W. A. Brock. Scaling in Economics: A Reader's Guide. *Industrial and Corporate Change*, 8:409–446, 1999.
29. W. A. Brock and S. N. Durlauf. Social Interactions and Macroeconomics, UW-Madison, SSRI Working Papers, 2005.
30. W. A. Brock and S. N. Durlauf. Discrete Choice with Social Interactions. *Review of Economic Studies*, 68:235–260, 2001.
31. S. Buldyrev, L. Amaral, S. Havlin, H. Leschhorn, P. Maass, M. Salinger, H. Stanley, and M. Stanley. Scaling Behavior in Economics: II. Modeling of Company Growth. *J. Phys. France I*, 7(635), 1997.
32. A. Burns and W. Mitchell. *Measuring Business Cycless.* New York, National Bureau of Economic Research, 1946.
33. R. J. Caballero. A Fallacy of Composition. *American Economic Review*, 1992.
34. D. Canning, L. A. N. Amaral, Y. Lee, M. Meyer, and H. E. Stanley. Scaling the Volatility of GDP Growth Rates. *Economic Letters*, 60:335–341, 1998.

35. C. D. Carroll. Requiem for the Representative Consumer? Aggregate Implications of Microeconomic Consumption Behavior. American Economic Review 90.2, 2000.
36. R. E. Caves. Industrial Organisation and New Findings on the Turnover and Mobility of Firms'. *Journal of Economic Literature*, 36:1947–1982, 1998.
37. H. L. Cole, G. J. Mailath, and A. Postlewaite. Social Norms, Savings Behavior, and Growth. *Journal of Political Economy*, 100(6):1092–1125, 1992.
38. R. Cont, M. Potters, and J.-P. Bouchad. *Scaling in Stock Market Data: Stable laws and Beyond*, in B. Dubrulle, F. Graner, and D. Sornette (eds.), Scale Invariance and Beyond. EDP Science, Les Ulis, 1997.
39. W. Cook and P. Ormerod. Power Law Distribution of the Demise of Firms in the US, *Physica A*, 2003.
40. R. Cooper, Coordination Games: Complementarities and Macroeconomics, Cambridge University Press, 1999.
41. K. Cowling and M. Waterson. Price-cost margins and market structure. *Economica*, 43(171):267–274, August 1976. Available at http://ideas.repec.org/a/bla/econom/v43y1976i171p267-74.html.
42. J. Crutchfield. Is Anything Ever New? Considering Emergence. In G. Cowan, D. Pines, and D. Meltzer, editors, *Complexity: Metaphors, Models, and Reality*, pp. 515–537. Reading, MA: Addison-Wesley, 1994. Available at http://www.santafe.edu/ jpc.
43. A. Czirok, H. E. Stanley, and T. Vicsek. Possible Origin of Power-Law Behavior in n-Tuple Zipf Analysis. *Phys. Rev. E*, 53:6371, 1996.
44. R. H. Day. *Complex Economic Dynamics, Vol. 1.* MIT Press, Cambridge, 1994.
45. S. Davis, Haltiwanger J. and S. Schub. *Job Creation and Destruction*, MIT Press, Cambridge, 1996.
46. A. T. Denzau and D. C. North. Shared Mental Models: Ideologies and Institutions. *Kyklos, Blackwell Publishing*, 47(1):3–31, 1994.
47. F. Diebold and G. Rudebusch. A Nonparametric Investigation of Duration Dependence in the American Business Cycles. *Journal of Political Economy*, 98(3):596–616, 1990.
48. F. Diebold, G. Rudebusch, and D. Sichel. Further Evidence on Business Cycle Duration Dependence. In: J. Stock and M. Watson (eds.), *Business Cycles, Indicators and Forecasting*. University of Chicago Press Chicago, IL, 1993.
49. D. Delli Gatti, C. Di Guilmi, and M. Gallegati, On the Empirics of "Bad Debt", mimeo, 2003.
50. D. Delli Gatti, M. Gallegati, and A. Kirman (eds.). Interaction and Market Structure. Essays on Heterogeneity in Economics, Lecture Notes in Economics and Mathematical Systems 484, Springer Verlag, Berlin, 2000.
51. D. Delli Gatti, C. Di Guilmi, E. Gaffeo, M. Gallegati, G. Giulioni, A. Palestrini. Business Cycle Fluctuations and Firms' Size Distribution Dynamics, in *Advances in Complex Systems*, vol. 7, no. 2, pp. 1–18, 2004.
52. D. Delli Gatti, C. Di Guilmi, E. Gaffeo, M. Gallegati, G. Giulioni, A. Palestrini. A New Approach to Business Fluctuations: Heterogeneous Interacting Agents, Scaling Laws and Financial Fragility, in *Journal of Economic Behaviour and Organization*, 56(4):489–512, 2005.
53. H. De Wit. Firm Size Distributions. An Overview of Steady-State Distributions Resulting from Firm Dynamics Models, *International Journal of Industrial Organization*, 23:423–450, 2005.

54. J. Diaz-Gimenez, V. Quadrini, and J.-V. Rios-Rull. Dimensions of Inequality: Facts on The U.S. Distributions of Earnings, Income and Wealth. *Federal Reserve Bank of Minneapolis Quarterly Review*, 21(2):3–21, 1997.
55. C. Di Guilmi, E. Gaffeo, and M. Gallegati. Empirical Results on the Size Distribution of Business Cycle Phases, mimeo, 2003.
56. E. Dinopoulos and P. Thompson. Schumpeterian Growth Without Scale Effects. *Journal of Economic Growth*, 3:313–335, 1998.
57. A. Dragulescu and V. Yakovenko. Exponential and Power-Law Probability Distributions of Wealth and Income in the United Kingdom and the United States. *Physica A*, 299:213–221, 2001.
58. B. Dupor. Aggregation and Irrelevance in Multi-Sector Models. *Journal of Monetary Economics*, 43(2):391–409, April 1999. Available at http://ideas.repec.org/a/eee/moneco/v43y1999i2p391-409.html.
59. S. N. Durlauf. Nonergodic Economic Growth. *Rewiew of Economic Studies*, 60:349–366, April 1993.
60. S. Durlauf. An Incomplete Markets Theory of Business Cycle Fluctuations. *Computational and Mathematical Organization Theory*, 2:191–212, 1996.
61. S. N. Durlauf. Complexity and Empirical Economics, mimeo, 2003.
62. S. N. Durlauf and H. P. Young. *Social Dynamics.* Cambridge, Mass: MIT Press, 2001.
63. K. Eng, X. G. Feng, D. Popovi, and S. Washburn. Effects of a Parallel Magnetic Field on the Metal-Insulator Transition in a Dilute Two-dimensional Electron System. *Phys. Rev. Lett.*, 88(13):136402, April 2002.
64. J. M. Epstein and R. Axtell. Growing Artificial Societies. Brookings Institution Press, Washington D.C., 1996.
65. R. Ericson and A. Pakes. Markov-Perfect Industry Dynamics: A Framework for Empirical Work. *Rev. Econ. Stud.*, 62:53, 1995.
66. P. Evans. Using Panel Data to Evaluate Growth Theory. *International Economic Review*, 39:259–306, 1998.
67. S. Fazzari, G. Hubbard, and B. Peterson. Financing Constraints and Corporate Investment. *Brooking Papers on Economic Activity*, 1:141–206, 1988.
68. W. Feller, An Introduction to Probability Theory and its Applications. Wiley, New York, 1957.
69. M. Forni, M. Lippi. Aggregation and the Microfoundations of Dynamic Macroeconomics, Oxford, Clarendon Press, 1997.
70. M. Friedman. The Role of Monetary Policy. *American Economy Rewiew*, 58:1–17, March 1968.
71. R. Frisch. *Propagation Problems and Impulse Problems in Dynamic Economics.* In: Essays in Honor of Gustav Cassel, 1933.
72. Y. Fujiwara, W. Souma, H. Aoyama, T. Kaizoji, and M. Aoki. Growth and Fluctuations of Personal Income *Physica A*, 321:598, 2003.
73. X. Gabaix. Zipf's Law for Cities: An Explanation. *Quarterly Journal of Economics*, 144:739–767, 1999.
74. X. Gabaix. The Granular Origins of Aggregate Fluctuations, MIT and NBER Working Paper, 2005.
75. E. Gaffeo, M. Gallegati, and A. Palestrini. On the Size Distribution of Firms, Additional Evidence from the G7 Countries, *Physica A*, 324:117–123, 2003.
76. M. Gallegati and L. Stanca. Financial Fragility, Heterogeneous Agents, and Aggregate Fluctuations: Evidence from a Panel of U.S. Firms. *Industrial and Corporate Change*, 1998.

77. M. Gallegati, Giulioni G., and N. Kichiji. Complex Dynamics and Financial Fragility in an Agent Based Model, *Advances in Complex Systems*, 2003.
78. M. Gallegati, A. Palestrini, D. Delli Gatti, and E. Scalas. Aggregation of Heterogeneous Interacting Agents: The Variant Representative Agent Framework, *Journal of Economic Interaction and Coordination*, 1(1):5–19, 2006.
79. R. Gibrat. *Les inégalités économiques.* Sirey, Paris, 1931.
80. E. Glaeser, B. Sacerdote, and J. Scheinkman. Crime and Social Interactions. *Quarterly Journal of Economics*, 111:507–548, 1996.
81. J. Glaser, J. Dixit, and D. P. Green. Studying Hate Crime with the Internet: What Makes Racists Advocate Racial Violence? *Journal of Social Issues*, 58:177–194, 2002.
82. B. Gnedenko and A. Kolmogorov. *Limit Distributions for Sums of Independent Random Variables.* Addison-Wesley, Reading, 1954.
83. R. Goodwin. A Growth Cycle. In: C. H. Feinstein (ed.), *Socialism, Capitalism and Economic Growth.* Cambridge University Press, 1967.
84. W. M. Gorman. Community Preference Fields. *Econometrica*, 21:63–80, 1953.
85. B. C. Greenwald and J. Stiglitz. *Macroeconomic Models with Equity and Credit Rationing*, in R. Hubbard (ed.), Information, Capital Markets and Investment. Chicago University Press, Chicago, 1990.
86. B. C. Greenwald and J. E. Stiglitz. Financial Market Imperfections and Business Cycles. *Quarterly Journal of Economics*, 108:77–114, 1993.
87. S. J. Grossman, J. E. Stiglitz On the Impossibility of Informationally Efficient Markets *American Economic Review*, 70(3):393–408, 1980.
88. J. C. Haltiwanger. Measuring and Analyzing Aggregate Fluctuations: The Importance of Building from Microeconomic Evidence. *Review*, pp. 55–78, 1997. Available at http://ideas.repec.org/a/fip/fedlrv/y1997imayp55-78.html.
89. F. Hahn, *Money and Inflation.* Oxford: Blackwell, 1982.
90. W. Härdle. Applied Nonparametric Regression. Cambridge University Press, Cambridge, 1990.
91. L. Hannah. Marshall's "Trees" and the Global "Forest": Were "Giant Redwoods" Different? In N. R. Lamoreaux, D. M. G. Raff and P. Temin (eds.), *Learning by Doing in Markets, Firms and Countries*, National Bureau of Economic Research, 1999.
92. D. Harding and A. Pagan. Dissecting the Cycle: A Methodological Investigation. *Journal of Monetary Economics, Elsevier*, 49(2):365–381, March 2002.
93. F. A. Hayek. *Individualisn and Economic Order,* University of Chicago Press, Chicago, 1948.
94. W. Hildenbrand and A. P. Kirman. *Equilibrium Analysis: Variations on the Themes by Edgeworth and Walras*, North-Holland, Amsterdam, 1988.
95. R. J. Hodrick and E. C. Prescott. Postwar U.S. Business Cycles: An Empirical Investigation. *Journal of Money, Credit and Banking*, 29(1):1–16, 1997. Available at http://ideas.repec.org/a/mcb/jmoncb/v29y1997i1p1-16.html.
96. C. H. Hommes. *Heterogeneous Agent Models in Economics and Finance*, In L. Tesfatsion, K. L. Judd (eds.), Handbook of Computational Economics, Volume 2: Agent-Based Computational Economics, Elsevier Science B.V., 2006.
97. C. H. Hommes. Partial Equilibrium Analysis in a Noisy Chaotic Market, *Economics Letters*, 53(3):275–282, 1996.

98. C. H. Hommes, and S. Manzan. Comments on "Testing for Nonlinear Structure and Chaos in Economic Time Series", *Journal of Macroeconomics* 28:169–174, 2006.
99. H. Hopenhayn. Entry, Exit and Firm Dynamics in Long-Run Equilibrium, *Econometrica*, 60:1127–1150, 1992.
100. G. Hubbard, K. Kuttner, and D. Palia. Are There "Bank Effects" in Borrowers' Costs of Funds?: Evidence from a Matched Sample of Borrowers and Banks, *Journal of Business*, 75:559–581, 2002.
101. C. R. Hulten, E. R. Dean, and M. J. Harper (eds.). *New Developments in Productivity. Analysis.* National Bureau of Economic Research Studies in Income and Wealth, 2001.
102. Y. Ijiri and H. Simon. *Skew Distributions and the Sizes of Business Firms.* North-Holland, Amsterdam, 1977.
103. IMF. *World Economic Outlook*, Washington D.C., 2001. October, Chapter II.
104. IMF. *World Economic Outlook*, Washington D.C., 2002.
105. Y. G. Joh, R. Orbach, G. G. Wood, J. Hammann, and E. Vincent. Extraction of the Spin Glass Correlation Length. *Phys. Rev. Lett.*, 82(2):438–441, Jan 1999.
106. C. Jones. On the Evolution of the World Income Distribution. *Journal of Economic Perspectives*, 11:19–36, 1997.
107. C. Jones. Growth: With or Without Scale Effects? *American Economic Review*, 89:139–144, 1999.
108. B. Jovanovic. Selection and the Evolution of Industry. *Econometrica*, 50(3):649–670, May 1982.
109. N. Kaldor. A Model of the Trade Cycle. *Economic Journal*, 50:78–92, 1940.
110. N. Kaldor. Capital Accumulation and Economic Growth. In F. Lutz and D. C. Haugue (eds.), *Proceedings of a Conference Held by the International Economics Association.* London, Macmillan, 1963.
111. H. H. Kalejan. *Aggregation and Disaggregation of Nonlinear Equations*, in J. Kmenta and J. B. Ramsey (eds.), Evaluation of Econometrics Models, Academic Press, New York, 1980.
112. M. Keeling. Simple Stochastic Models and Their Power-Law Type Behaviour Theo. *Pop. Biol.*, 58:21–31, 2000.
113. M. Keeling and B. Grenfell. Stochastic Dynamics and a Power Law for Measles Variability, *Philosophical Transactions of the Royal Society B (Biological Sciences)*, 354:769–776, 1999.
114. T. Keitt and H. Stanley. Dynamics of North American Breeding Bird Populations. *Nature*, 393:257–260, 1998.
115. W. J. Keller. Aggregation of Economic Relation: A Note, *De Economist*, 128:559–562, 1984.
116. H. Kesten. Random Difference Equations and Renewal Theory for Products of Random Matrices. Acta Mathematica, CXXXI: 207–248, 1973.
117. F. E. Kidland and E. C. Prescott. Time to Build and Aggregate Fluctuations. *Econometrica*, 50:1345–1370, 1982.
118. F. E. Kidland and E. C. Prescott. Business Cycles: Real Facts and a Monetary Myth. *Federal Reserve Bank of Minneapolis, Quarterly Review*, 14:318, 1990.
119. A. Kilpatrick and A. Ives. Species Interactions Can Explain Taylor's Power Law for Ecological Time Series. *Nature*, 422:65–68, March 2003.
120. A. P. Kirman. The Intrinsic Limits of Modern Economic Theory: The Emperor Has No Clothes. *Economic Journal*, 99(395):126–139, 1989.

121. A. P. Kirman. Whom or What Does the Representative Individual Represent. *Journal of Economic Perspective*, 6:117–136, 1992.
122. T. Knudsen. Zipf's Law for Cities and Beyond, the Case of Denmark. *Am. J. Econ. Sociol.*, 60:123–146, 2001.
123. R. Kroszner and P. E. Strhan. What Drives Deregulation? Economics and Politics of the Relaxation of Bank Branching Restrictions. Quarterly Journal of Economics, 114(4):1437–1467, 1999.
124. J. J. Kruger. On the Dynamics of the U.S. Manufacturing Productivity Distribution. Working paper, Friedrich-Schiller Universität Jena, 2003.
125. P. Krugman. *The Self-Organizing Economy.* Cambridge, Blackwell, 1996.
126. J. E. Kwoka, Regularity and Diversity in Firm Size Distributions in U.S. Industries, *Journal of Economics and Business*, 34:391–395, 1982.
127. T. Lancaster. *The Econometric Analysis of Transition Data.* Cambridge University Press, Cambridge, 1992.
128. T. Lancaster. Econometric Methods for the Duration of Unemployment. *Econometrica*, 47:939–956, 1979.
129. Y. Lee, L. Amaral, D. Canning, M. Meyer, and H. Stanley. Universal Features in the Growth Dynamics of Complex Organizations. *Physical Review Letters*, 81:3275–3278, 1998.
130. H. Leibenstein. Bandwagon, Snob, and Veblen Effects in the Theory of Consumers' Demand. *Quarterly Journal of Economics*, 64:183–207, May 1950.
131. A. Lewbel. Exact Aggregation and a Representative Consumer, *Quarterly Journal of Economics*, 104(3):621–633, 1989.
132. J. Long and C. Plosser. Real Business Cycles. *Journal of Political Economy*, 91(1):39–69, 1983.
133. R. E. J. Lucas. An Equilibrium Model of the Business Cycle. *Journal of Political Economy*, 83:1113–1144, December 1975.
134. R. E. J. Lucas. *Studies in Business-Cycle Theory.* MIT Press, 1981.
135. Y. Malavergne, V. Pisaneko, and D. Sornette. Empirical Distributions of Log-Returns: Between the Stretched Exponential and the Power Law? Working Paper, 2003. Available at http://papers.ssrn.com/sol3/paper.cfm.
136. O. B. Malcai and S. Solomon. Power-Law Distributions and Lévy-Stable Intermittent Fluctuations in Stochastic Systems of Many Autocatalytic Elements. *Phys. Rev. E*, 60:1299, 1999.
137. B. Mandelbrot. The Pareto-Levy Law and the Distribution of Income. *International Economic Review*, I, pp. 79–106, 1960.
138. C. F. Manski. Economic Analysis of Social Interactions. *Journal of Economic Perspectives*, 14:115–136, 2000.
139. R. Mantel. Homothetic Preferences and Community Excess Demand Functions, *Journal of Economic Theory* 12:197–201, 1976.
140. Marshall. *Principles of Economics.* Macmillan and Co., Ltd., London, 1920. 8th edition.
141. A. W. Marshall and I. Olkin. A Multivariate Exponential Distribution. Journal of American Statistical Association, 1967.
142. M. Marsili and Y.-C. Zhang. Interacting Individuals Leading to Zips' Law, *Physical Review Letters*, 80:2741–2744, 1998.
143. J. McCauley. *Dynamics of Markets, Econophysics and Finance.* Cambridge University Press, Cambridge, 2004.
144. H. J. McCulloch. The Monte Carlo Cycle in Economic Activity. *Economic Inquiry*, 13:303–321, September 1975.

145. J. McDonald and R. J. Butler. Some Generalized Mixture Distributions with an Application to Unemployment Duration. *The Review of Economics and Statistics, MIT Press*, 69(2):232–240, May 1987.
146. P. Mirowski. *More Heat than Light.* Cambridge University Press, Cambridge, 1989.
147. M. Mizuno, M. Katori, H. Takayasu, and M. Takayasu. Statistical and Stochastic Laws in the Income of Japanese Companies. In H. Takayasu (ed.), *Empirical Science of Financial Fluctuations: The Advent of Econophysics*, p. 321, Springer, Tokyo, 2002.
148. E. W. Montroll and M. F. Shlesinger. Maximum Entropy Formalism, Fractals, Scaling Phenomena, and $1/f$ Noise: A Tale of Tails. *J. Stat. Phys.*, 32(2):209–30, 1983.
149. G. Mudholkar, D. K. Srivastava, and G. D. Kollia. A Generalization of the Weibull Distribution with Application to the Analysis of Survival Data. *Journal of the American Statistical Association*, 91:1575–1583, 1996.
150. M. Naldi. Concentration Indices and Zipf's Law. *Economics Letters*, 78(3), 2003.
151. M. Nirei and W. Souma. Two Factors Model of Income Distribution Dynamics. Santa Fe Institute Working Paper, 04-10-029, 2004.
152. J. Nolan. *Stable Distributions.* Springer, Berlin, 2002.
153. S. Nørrelykke and P. Bak. Self-Organized Criticality in a Transient System, *Physics Review*, 68, 2002.
154. A. Palestrini. Analysis of Industrial Dynamics: a Note on the Relationship Between Firms' Size and Growth Rate, *Economics Letters*, 94:367–371, 2007.
155. V. Pareto. *Le Cours d'Économie Politique.* Macmillan, London, 1897.
156. M. H. Pesaran. On Aggregation of Linear Dynamic Models, mimeo, 2000.
157. P. Perron. The Great Crash, the Oil Price Shock, and the Unit Root Hypothesis, *Econometrica*, 57:1361–1401, 1989.
158. H. Platt. *Why Companies Fail.* D.C. Heath and Company Lexington, 1985.
159. C. I. Plosser. Understanding Real Business Cycles. *Journal of Economic Perspectives, American Economic Association*, 3(3):51–77, Summer 1989.
160. R. Pollack. Interdependent Preferences, *American Economic Review*, 66:309–320, 1975.
161. D. M. W. Powers. Applications and Explanations of Zipf's Law. In: J. Burstein and C. Leacock (eds.), *Proceedings of the Joint Conference on New Methods in Language Processing and Computational Language Learning*, Association for Computational Linguistics, Somerset, New Jersey, 151–160, 1998.
162. L. Pritchett. Divergence, Big Time. *Journal of Economic Perspectives*, 11:3–17, 1997.
163. D. Quah. Empirical Cross-Section Dynamics in Economic Growth. *European Economic Review*, 37:426–434, 1993.
164. D. Quah. Twin Peaks: Growth and Convergence in Models of Distribution Dynamics. *Economic Journal*, 106:1045–1055, 1996.
165. R. E. Quandt. On the Size Distribution of Firms, *American Economic Review*, 56:416–432, 1966.
166. W. Reed. The Double Pareto – Lognormal Distribution a New Parametric Model for Size Distributions. unpublished manuscript, http://www.math.uvic.ca/faculty/reed/., 2001.

167. S. Rebelo. Long-Run Policy Analysis and Long-Run Growth, *Journal of Political Economy*, 99:500–521, 1991.
168. S. Resnik. *Extreme Values, Regular Variation, and Point Processes.* New York: Springer Verlag, 1987.
169. C. D. Romer. Changes in Business Cycles: Evidence and Explanations. *Journal of Economic Perspectives*, 13:23–44, Spring 1999.
170. G. Samorodnitsky and M. Taqqu. Stable Non-Gaussian Random Processes, Chapman and Hall. 1994
171. T. C. Shelling. *Micromotives and Macrobehavior*, W. W. Northon, NY, 1978.
172. R. Schmalensee. Sunk Costs and Market Structure: A Review Article. *J. Ind. Econ.*, 40:125–133, 1989.
173. J. Schumpeter. *Business Cycles.* McGraw-Hill, New York, 1939.
174. P. Segerstrom. Endogenous Growth Without Scale Effect. *American Economic Review*, 88:1290–1310, 1998.
175. O. Shy. *The Economics of Network Industries.* Cambridge University Press, Cambridge, 2001.
176. D. Sichel. Business Cycle Duration Dependence: A Parametric Approach. *Review of Economics and Statistics*, 71(30):245–260, 1991.
177. D. E. Sichel. Business Cycle Asymmetry: A Deeper Look. *Economic Inquiry*, 31:224–236, April 1993.
178. J. Siegfried and L. Evans. Empirical Studies of Entry and Exit: A Survey of the Evidence. *Review of Industrial Organization*, 9:121–155, 1994.
179. H. Simon. On a Class of Skew Distribution Functions, *Biometrika*, 27:425–440, 1955.
180. R. Sinclair. Examining the Growth Model's Implications: The World Income Distribution. Syracuse University, mimeo, 2001.
181. E. Slutzky. The Summation of Random Causes as the Source of Cyclical Processes. *Econometrica*, 5(2):105–146, 1937.
182. S. Smale. Dynamics in General Equilibrium Theory. *American Economic Review*, 66(2):284–294, May 1976.
183. A. Smith. *An Inquiry into the Nature and Causes of the Wealth of Nations*, American Modern Library Series, NY, Cannan Edition, 1937.
184. H. F. Sonnenschein. Market Excess Demand Functions, *Econometrica*, 1972.
185. D. Sornette. *Critical Phenomena in Natural Sciences, Chaos, Fractals, Self-Organization and Disorder. Concepts and Tools.* Springer, Heidelberg, 2000.
186. W. Souma. Universal Structure of the Personal Income Distribution. *Fractals*, 9(4):463–470, 2001.
187. M. Stanley, S. Buldyrev, S. Havlin, R. Mantegna, M. Salinger, and E. Stanley. Zipf Plots and the Size Distribution of Firms. *Economics Letters*, 49:453–457, 1995.
188. M. Stanley, L. Amaral, S. Buldyrev, S. Havlin, H. Leschorn, P. Maas, M. Salinger, and E. Stanley. Scaling Behavior in the Growth of Companies, *Nature*, 379:804–806, 1996.
189. J. Steindl. *Random Processes and the Growth of Firms.* Hafner, New York, 1965.
190. J. H. Stock and M. W. Watson. Understanding Changes in International Business Cycle Dynamics. *Journal of the European Economic Association MIT Press*, 3(5):968–1006, 2005.
191. T. M. Stoker. Completeness, Distribution Restriction, and the Form of Aggregate Functions, *Econometrica*, 52:887–907, 1984.

192. T. M. Stoker. Empirical Approaches to the Problem of Aggregation Over Individuals, *Journal of Economic Literature*, 31(4):1827–1874, 1993.
193. R. Summers, A. Heston, and B. Aten. Penn World Table Version 6.1. Center for International Comparisons at the University of Pennsylvania, 2002.
194. S. Sunder. Determinants of Economic Interaction: Behavior and Structure. *Journal of Economic Interaction and Coordination*, 1:21–32, 2006.
195. J. Sutton. Gibrat's Legacy. Journal of Economic Literature 35, 40–59, 1997.
196. J. Sutton. *Technology and Market Structure. Theory and History.* MIT Press, 1999.
197. H. Takayasu and K. Okuyama. Country Dependence on Company Size Distributions and a Numerical Model Based on Competition and Cooperation. *Fractals*, 6:67–79, 1998.
198. J. B. Taylor. Aggregate Dynamics and Staggered Contracts. *Journal of Political Economy*, 88:1–23, February 1980.
199. L. Taylor. Aggregation, Variance and the Mean. *Nature 189*, pp. 732–735, 1961.
200. L. Taylor and I. Woiwod. Comparative Synoptic Dynamics. I. Relationships Between Interspecific and Intraspecific Spatial and Temporal Variance-Mean Population Parameters. *J. Anim. Ecol.*, 51:879–906, 1982.
201. L. Taylor, I. Woiwod, and J. Perry. The Density-Dependence of Spatial Behaviour and the Rarity of Randomness. *J. Anim. Ecol.*, 47:383–406, 1978.
202. L. Tesfatsion. Agent-based Computational Economics: Growing Economies from the Bottom-Up. ISU Economics Working Paper No. 1, February 2002.
203. L. Tesfatsion, K. L. Judd (eds.). *Handbook of Computational Economics, Volume 2: Agent-based Computational Economics*, Elsevier Science B.V., 2006.
204. H. Theil. Linear Aggregation of Economic Relations. Amsterdam, North-Holland Publishing Company, 1954.
205. J. Voit. The Growth Dynamics of German Business Firms. cond-mat/0006260 v1, 2000.
206. S. Winter, Y. Kaniovski, and G. Dosi. A Baseline Model of Industry Evolution. Interim Report IR-97-013/March, International Institute for Applied Systems Analysis, Laxenburg, Austria, 1997.
207. D. Zajdenweber. *Scale Invariance in Economics and in Finance.* In: B. Dubrulle, F. Graner, and D. Sornette (eds.), Scale Invariance and Beyond. EDP Science, Les Ulis, 1997.
208. V. Zarnowitz. *Business Cycles: Theory, History, Indicators, and Forecasting.* The University of Chicago Press, Chicago, 1992.
209. T. W. Zuehlke. Business Cycle Duration Dependence Reconsidered. *Journal of Business and Economic Statistics, American Statistical Association*, 21(4):564–569, October 2003.

Printed in March 2008

MIX
Papier aus verantwortungsvollen Quellen
Paper from responsible sources
FSC® C105338

If you have any concerns about our products,
you can contact us on
ProductSafety@springernature.com

In case Publisher is established outside the EU,
the EU authorized representative is:
Springer Nature Customer Service Center GmbH
Europaplatz 3, 69115 Heidelberg, Germany

Printed by Libri Plureos GmbH
in Hamburg, Germany